The aim of this book series is to provide a platform to build and consolidate a rigorous and significant repository of academic, practice and research publications that contribute to further knowledge relating to management in the built environment. Its objectives are to:

(1) Disseminate new and contemporary knowledge relating to research and practice in the built environment
(2) Promote synergy across different research and practice domains in the built environment and
(3) Advance cutting-edge research and best practice in the built environment

The scope of this book series is not limited to "management" issues per se because this then begs the question of what exactly are we managing in the built environment. While the primary focus is on management issues in the building and construction industry, its scope has been extended upstream to the design management phase and downstream to the post-occupancy facilities management phase. Management in the built environment also involves other closely allied disciplines in the areas of economics, environment, legal and technology.Hence, the starting point of this book series lieswith project management, extends into construction and ends with facilities management. In between this spectrum, there are also other management-related issues that are allied with or relevant to the built environment. These can include, for example cost management, disaster management, contract management and management of technology.

This book series serves to engage and encourage the generation of new knowledge in these areas and to offer a publishing platform within which different strands of management in the built environment can be positioned to promote synergistic collaboration at their interfaces. This book series also provides a platform for other authors to benchmark their thoughts to identify innovative ideas that they can further build on to further advance cutting-edge research and best practice in the built environment.

If you are interested in submitting a proposal for this series, please kindly contact the Series Editor or the Publishing Editor at Springer:

Low Sui Pheng (bdglowsp@nus.edu.sg) or
Ramesh Premnath (Ramesh.premnath@springer.com)

More information about this series at http://www.springer.com/series/15765

Management in the Built Environment

Series editor

Low Sui Pheng, National University of Singapore, Singapore, Singapo

Low Sui Pheng · Sim Jia Zheng

Construction Productivity in the Multilayer Subcontracting System

The Case of Singapore

Springer

Low Sui Pheng
Department of Building
National University of Singapore
Singapore, Singapore

Sim Jia Zheng
Department of Building
National University of Singapore
Singapore, Singapore

ISSN 2522-0047 ISSN 2522-0055 (electronic)
Management in the Built Environment
ISBN 978-981-13-1830-6 ISBN 978-981-13-1831-3 (eBook)
https://doi.org/10.1007/978-981-13-1831-3

Library of Congress Control Number: 2018950786

This Springer imprint is published by the registered company Springer Nature Singapore Pte Ltd.
The registered company address is: 152 Beach Road, #21-01/04 Gateway East, Singapore 189721, Singapore

Contents

Abbreviations

BCA	Building and Construction Authority
BIM	Building Information Modelling
CPCF	Construction Productivity and Capability Fund
DRC	Dependency Ratio Ceiling
FWL	Foreign Worker Levy
GDP	Gross Domestic Product
HR	Human Resource
ILO	International Labour Organisation
KPI	Key Performance Indicator
MC	Main Contractor
M&E	Mechanical and Electrical
MLSS	Multi-Layered Subcontracting System
MYE	Man-Year Entitlement
NSC	Nominated Subcontractor
PM	Project Manager
P&S	Planning and Scheduling
QDA	Qualitative Data Analysis
QS	Quantity Surveyor
RBV	Resource-Based View
RE	Resident Engineer
RTO	Resident Technical Officer
SC	Subcontractor
SCA	Sustained Competitive Advantage
SME	Small and Medium Enterprises
SPSS	Statistical Package for Social Sciences
UK	United Kingdom

List of Figures

List of Tables

Chapter 1
Introduction

Abstract This chapter presents the general background behind this research and the introduction of the research problem. This chapter states the purpose of this study that is to explore and better understand how the fragmentation of large and small firms in the Singapore construction industry affects productivity in the multilayer subcontracting system (MLSS).

1.1 Background

Fragmentation has long been existent in the construction industry and has resulted in its poor performance (Egan, 1998). It is widely debated as a contributing factor that has caused the industry to be inclined towards low productivity, poor value for money, and mediocre overall client satisfaction when compared to other industry sectors (Latham, 1994). Moreover, the inherent difficulty and uncertainty for main contractors in obtaining continuous work and accommodating increasingly diverse features and requirements of projects have resulted in the adoption of subcontracting practices as the dominant approach, which has caused further fragmentation due to its very nature (Cox & Townsend, 1998).

At present, as there is a high tendency for the main contractor to engage subcontractors to meet their labour and/or material requirements (Loh & Ofori, 2000), this has further exacerbated the effects of fragmentation in the local industry. Due to the uniqueness of each construction project in which multiple crafts are involved and the limited timeframes and variety of materials and equipment required, main contractors tend to outsource jobs to subcontracting firms that provide temporary manpower as opposed to directly hiring construction workers. Such a practice has given rise to the adoption of 'flexible' labour practices where indirect employment occurs, in which the main contractor tends to outsource work to subcontractors to fulfil labour requirements (ILO, 2001). In this paper, a subcontractor is defined as a construction firm that enters into a contract or several contracts with a main contractor to perform some of the main contractor's workload. Basic operations will be performed by the main contractor, while the remaining work will be subcontracted

L. Sui Pheng and S. Jia Zheng, *Construction Productivity in the Multilayer Subcontracting System*, Management in the Built Environment,
https://doi.org/10.1007/978-981-13-1831-3_1

to various subcontractors depending on their specialisations (Lew, Hassim, Muniandy, & Law, 2012). In addition, the subcontracting phenomenon is not only seen in Singapore but also in construction industries of other states.

However, the local construction industry has often been singled out for lagging behind many other industries, despite the fact that the industry contributes a significant 5% to Singapore's gross domestic product (GDP) as reflected in Table 1.1. A recent statistic has shown that the productivity growth of Singapore's construction industry is less than that of other local industries, with a 2.9% shrinkage in 2013 (Low, 2015), and at the halfway mark in 2015, labour productivity fell into negative territory, contracting 0.8% last year (Chia, 2015). All these occurred despite the government's efforts in allocating one quarter ($250 million) of the $1 billion National Productivity Fund through the Construction Productivity and Capability Fund (CPCF) to help construction firms in three key areas: manpower development and skills upgrading, technology adoption, and capability building. This aims to achieve the target of 2–3% productivity growth per year for the next decade, where the construction industry identifies productivity-based growth opportunities that can help sustain Singapore's economy (Building and Construction Authority, 2010).

Table 1.1 Total economy and construction gross domestic product (GDP; at 2010 market prices)

Year	Total Economy ($m)	Construction ($m)	GDP contribution (%)
1994	130,251.0	6709.1	5.2
1995	139,405.1	7284.5	5.2
1996	149,905.0	8750.8	5.8
1997	162,333.8	9894.0	6.1
1998	158,721.5	10,094.0	6.4
1999	168,395.9	9002.5	5.3
2000	183,379.0	8887.6	4.8
2001	181,632.7	8906.8	4.9
2002	189,282.5	7767.5	4.1
2003	197,677.8	7122.8	3.6
2004	216,554.4	6758.1	3.1
2005	232,772.5	6847.3	2.9
2006	253,396.6	7298.4	2.9
2007	276,484.9	8757.9	3.2
2008	281,427.4	10,893.2	3.9
2009	279,729.3	13,233.2	4.7
2010	322,361.1	14,221.2	4.4
2011	342,371.8	14,914.5	4.4
2012	354,937.3	16,618.9	4.7
2013	371,531.5	17,578.1	4.7
2014	383,643.6	18,193.6	4.7
2015	391,348.5	18,640.3	4.8

Source Singapore Department of Statistics

1.2 Research Problem

The practice of subcontracting does provide certain economies of scale to the main contractor, as there is no need to hire permanent groups of workers or procure expensive machinery. It helps to relieve certain parts of the main contractor's work load and minimises the risks they face when it is shared with the subcontractors through a contractual arrangement. Moreover, as the main contractor is able to react effectively and efficiently to the fluctuations in construction demand, this eventually gives rise to a high volume of subcontracting.

Nonetheless, subcontracting has its downsides for both the main contractor and subcontractor. Poor skills and incompetent subcontractors and their workers were alleged to be the reason for the poor quality of construction work, which happens to be one of the key obstacles faced by main contractors (Yik, Lai, Chan, & Yiu, 2006a, b, c). As for subcontractors, poor treatment and delayed payment by the main contractor are common occurrences (CIDB, 2013). Despite such setbacks, the practice of subcontracting will not be going away anytime soon. According to Ofori and Lim (2009), Singapore's construction industry is established to have a MLSS and most construction firms in Singapore are subcontractors, consisting of small and medium enterprises (SMEs).

Taymaz and Kilicaslan (2002) proposed the notion that the SME industry would be eradicated by more efficient and advanced large firms because, similar to the local context, subcontracting has always been considered a system of the large firm's supremacy over smaller firms, as bigger firms reap the benefits of the low wages and flexible work arrangements in the more numerous smaller firms. It seemed to hold true until local research has proven that, in the Singapore context, SMEs have flourished and extended their influence, as they offer numerous economic benefits to their customers and business partners (Ofori & Lim, 2009). As such, the general notion that the fragmentation caused by having many small construction firms over large firms would result in the 'death' of SMEs when competing with their larger counterparts might have been misconstrued due to misconceptions.

In addition, it seems to be a timely question to ask about the ability of SME subcontractors and their larger main contractor counterparts to survive in an industry that adopts the MLSS, where there are considerable changes to many government policies and initiatives within the past five years (Heng, 2015), for instance, the introduction of stringent labour restructuring policies and government funding schemes like CPCF, etc.

After a review of several literature and study articles on the probable reasons for subcontracting, a research gap was identified in the Singapore context. There is little understanding of the multi-layered subcontracting system (MLSS) in the Singapore construction industry, particularly in the area of trade/specialist subcontracting and how such fragmentation has an effect on construction productivity in Singapore. Therefore, it warrants further investigation.

1.3 Research Objectives

The purpose of this study will serve to explore and better understand how the fragmentation of large and small firms in Singapore affects construction productivity when operating under the commonly adopted MLSS in the context of the local construction industry.

To achieve the purpose of this study, the specific objectives are as follows:

- To understand the nature of Singapore's construction industry to enable a comprehensive view of the workings of Singapore's construction industry.
- To assess the level of fragmentation in the Singapore construction industry through the use of statistical analysis.
- To examine the current status of subcontracting practices being implemented in the local industry to determine the degree of subcontracting being implemented by local contractors.
- To study the effectiveness of subcontracting practices in helping local contractors to improve their construction productivity and thereby improve the industry's productivity.

1.4 Research Scope

This research focuses on the current subcontracting practices in the construction industry and their effects on Singapore's construction productivity. In terms of relationships within the industry, there are three common classifications of contractors. Main contractors are those who successfully tendered for a project from the client/owner and liaise directly with them to meet their project needs. Subcontractors are those who work directly under the main contractors while secondary subcontractors are subcontractors who work under other subcontractors. The practice of subcontracting work by one subcontractor to another, regardless of the category or tier of subcontractor, is called secondary subcontracting, whereas primary subcontracting refers to the practice of subcontracting from the main contractor to subcontractors. As such, the effects of subcontracting practices, via subcontracting relationships, particularly among the main contractor and the trade and specialist subcontractors, on construction productivity will be examined without much emphasis on labour-only subcontracting since it has already been covered extensively in many research studies; see for example, An and Watanabe (2011), Ofori and Lim (2009), and Ofori and Yaw (1998).

1.5 Research Hypothesis

The fragmentation of the local construction industry that has caused many large and small firms to coexist and compete against each other has produced an MLSS where subcontracting practices have a significant effect on construction productivity in Singapore. It is to be concluded that firms that undertake subcontracting achieve higher work efficiency and therefore accrue increased productivity, thereby raising the overall industry's productivity.

1.6 Study Structure

This study comprises eight chapters. The following provides a brief outline of the chapters. This chapter, Introduction, highlights the general background, the rationale behind this research, and the structure of the study through the introduction of the research problem, objectives, scope, and hypothesis of this study. It also reviews previous studies with similar topics. Chapters 2 to 4 present the literature review of this research study. Chapter 2, Nature of the Construction Industry, provides insight on the characteristics of construction industries found globally and their commonalities. A discussion on subcontracting practices in the industry will also be addressed here. Chapter 3, Outlook on Singapore's Construction Industry, seeks to assess the level of fragmentation in the Singapore construction industry with the aid of BCA's data on 'Survey of the Construction Industry' that spans from 1997 to 2015. Chapter 4, Theoretical Framework, draws upon ideas and theories of past researchers on the reasons firms still adopt subcontracting practices in a fragmented industry via literature reviews of publications and journals. This will form the basis of understanding how subcontracting has an effect on industry-wide construction productivity. Chapter 5, Research Design and Methodology, presents the research methods adopted for this study and its design in order to attain the aims and objectives. Chapter 6, Analysis of Fieldwork, presents the interview findings and the analysis of the results. Chapter 7, Significance of Fieldwork Analysis, seeks to put the findings from the fieldwork analysis into perspective and make a relation to construction productivity. This is to exemplify the industry sentiment that subcontracting practices have a significant effect on construction productivity. Chapter 8, Effect on Construction Productivity, seeks to establish and elaborate on the link between industry fragmentation due to adoption of subcontracting practices and its effects on productivity based on the analysis of fieldwork results. Chapter 9, Conclusion, provides the recommendations, highlights the limitations in the research process, concludes the study, and suggests possible areas for further research. All references and sources can be found under the bibliography section, and the supporting details and information for this study can be found in the appendices.

Figure 1.1 illustrates the structure of this research study.

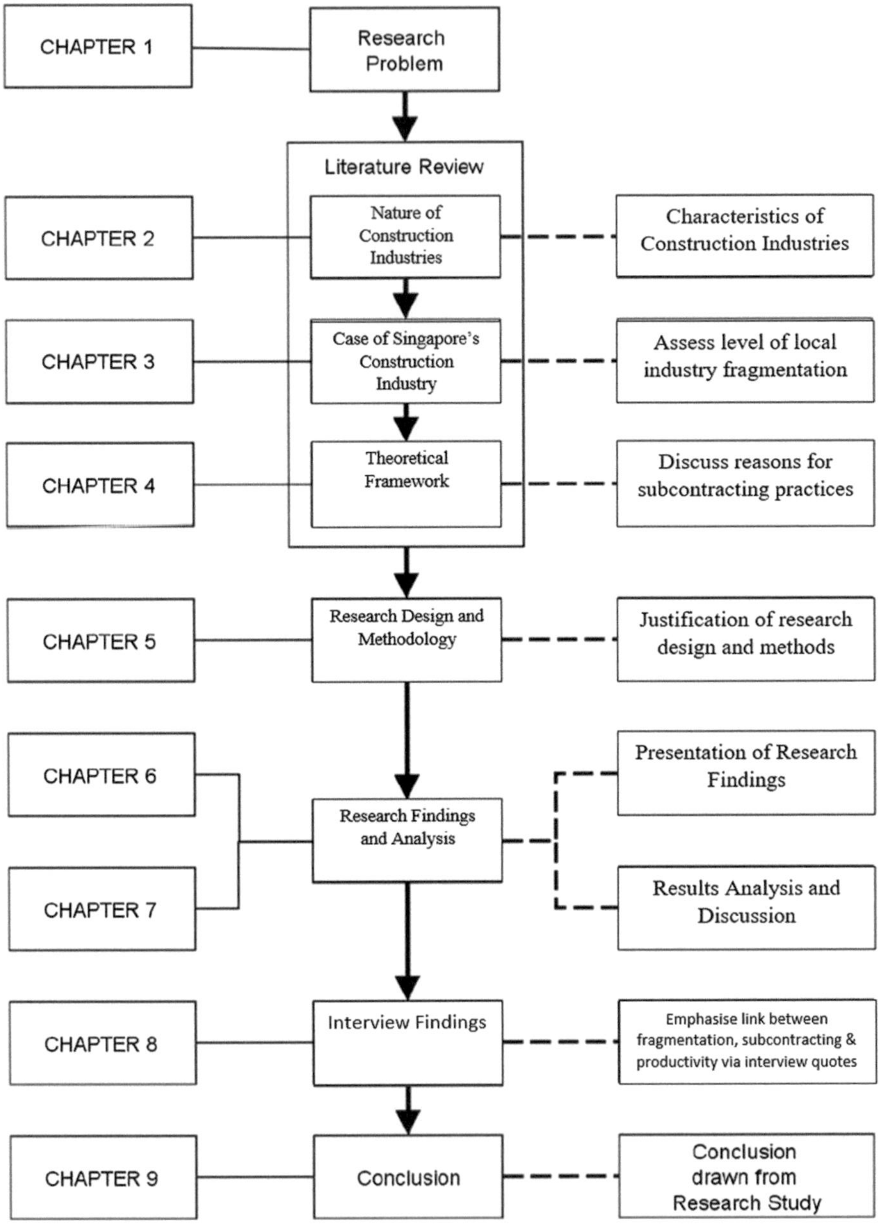

Fig. 1.1 Study structure

References

An, T., & Watanabe, T. (2011). Towards the sustainable improvement in construction labour's employment in China: Cultivating subcontractor's role. In C. Egbu & E. C. W. Lou (Eds.), *Proceedings 27th Annual ARCOM Conference* (pp. 663–672), September 5–7, 2011. Bristol, UK: Association of Researchers in Construction Management.

Building and Construction Authority. (2010). *BuildSmart*. Singapore, Issue no. 1:1.

Chia, Y. (2015). Labour productivity in Singapore contracts 0.8% in 2014, extending weak performance. *The Straits Times*. Retrieved September 16, 2016, from http://www.straitstimes.com/business/economy/labour-productivity-in-singapore-contracts-08-in-2014-extending-weak-performance.

CIDB. (2013). *Subcontracting in the South African construction industry; opportunities for development*. South Africa: Construction Industry Development Board. Retrieved April 18, 2014 from http://www.cidb.org.za/Documents/KC/cidb_Publications/Ind_Reps_Other/ind_reps_Subcontracting_in_the_SA_Construction_Industry.pdf.

Cox, A., & Townsend, M. (1998). *Strategic procurement in construction: Towards better practice in the management of construction supply chains*. London: Thomas Telford.

Egan, J. (1998). *Rethinking construction: The report of the construction task force*. London: HMSO.

Heng, N. (2015). *Labour restructuring policies and their implications on the multi-layered subcontracting system in the Singapore construction industry* (Unpublished BSc. Building study). National University of Singapore, Singapore

ILO. (2001). The construction industry in the twenty-first century: Its image, employment prospects and skill requirements. Switzerland: International Labour Office. Retrieved July 3, 2014 from http://www.voced.edu.au/content/ngv6183.

Latham, M. (1994). Constructing the team: Final report: *Joint review of procurement and contractual arrangements in the United Kingdom construction industry*. London: HMSO.

Lew, Y. L., Hassim, S., Muniandy, R., & Law, T. H. (2012). Review of subcontracting practice in construction industry. *IACSIT International Journal of Engineering and Technology, 4*(4), 442–445.

Loh, W. H., & Ofori, G. (2000). Effect of registration on performance of construction subcontractors in Singapore. *Engineering, Construction and Architectural Management, 7*(1), 29–40.

Low, C. (2015). *Productivity concepts and techniques* (Unpublished Lecture notes). Building and Construction Authority Academy, Singapore.

Ofori, G., & Debrah, Y. A. (1998). Flexible management of worker: Review of employment and practices in the construction industry in Singapore. *Construction Management and Economics,16(4)*, 397–408.

Ofori, G., & Lim, K. H. (2009). Multi-layered Subcontracting System in Singapore's Construction Industry. In K. Kobayashi, A. R. Kharuddin, G. Ofori, & S. Ogunlana (Eds.), *Joint Ventures in Construction* (pp. 216–226). Great Britain: Thomas Telford Limited.

Taymaz, E., & Kilicaslan, Y. (2002). *Subcontracting dynamics and economic development: A study on textile and engineering industries*. Economic Research Centre Working Papers in Economics 01/08 (pp. 1–25). Ankara, Turkey: Economic Research Centre, Middle East Technical University.

Yik, F., Lai, J., Chan, K. T., & Yiu, E. (2006a). *Best practices in managing specialist subcontracting performance*. Hong Kong: The Hong Kong Polytechnic University. Retrieved May 13, 2014 from http://repository.lib.polyu.edu.hk/jspui/bitstream/10397/2347/1/Full_Report_on_Best_Practices_in_Managing_Specialist_Subcontracting_Performance.pdfFresearch.

Yik, F., Lai, J., Chan, K., & Yiu, E. (2006b). *Best practices in managing specialist subcontracting practices* (Final Research Project Report). Hong Kong: Construction Industry Institute.

Yik, F. W. H., Lai, J. H. K., Chan, K. T., & Yiu, E. C. Y. (2006c). Problems with specialist subcontracting in the construction industry. *Building Services Engineering Research & Technology, 22*(1), 183–193.

Chapter 2
Nature of the Construction Industry

Abstract This chapter provides insight on the characteristics of construction industries found globally and their commonalities. It presents the subcontracting practices in the industry and highlights the high number of project participants, each with their distributed responsibilities, as they provide information, knowledge, and manpower resources during the construction process and deliver the final product to the client at the end of the process. This resulted in a complex network organisation arising from segregation in the industry. This chapter highlights the multi-cultural, multi-location, multi-disciplinary, and multi-organisational nature of the project participants.

2.1 General Features of the Construction Industry

The general perception of Singapore's construction industry at present has been described as dirty, demanding, and dangerous (Construction 21 Steering Committee, 1999). The industry also has a high number of various project participants, each with their distributed responsibilities, as they provide information, knowledge, and manpower resources during the construction process and deliver the final product to the client at the end of the process. As such, this has resulted in a complex network organisation as a result of the nature of the segregation in the industry, where there is a multi-cultural, multi-location, multi-disciplinary, multi-organisational nature of the project participants (Schwegler, Fischer, O'Connell, Hanninen, & Laitinen, 2001). Thus, this has further worsened the fragmentation, as these parties often do not come together for discussion because such coordination is not easy.

2.1.1 Effects of Fragmentation in Construction Industries

With global external events, such as the escalating cost of construction due to cost increases in equipment, labour, and materials the effects of fragmentation have

L. Sui Pheng and S. Jia Zheng, *Construction Productivity in the Multilayer Subcontracting System*, Management in the Built Environment,
https://doi.org/10.1007/978-981-13-1831-3_2

further worsened. Foo and Ho (2001) pointed out that the construction industry is subject to the business cycle. An economic downturn would aggravate price increases due to the collapse of several small suppliers; conversely, an economic upswing could result in exceptionally high demand in times of construction booms, causing further increases in cost. The effects of fragmentation can be felt when there is intense competition in the industry, where the players are sceptical of sharing information or resources, which causes the industry to become less trusting and more self-interested and adversarial. Adversarial relationships and fragmented processes significantly hinder the performance and innovation in construction (Morledge, Knight, & Grada, 2009). Therefore, all these factors contribute to the industry's fragmentation by the nature of its operation and the mechanism of the market forces where enterprises are often 'disconnected' from one another, operating independently with little or no synergy among them.

In addition, the Latham Report discussed various attributes of construction industries worldwide. For instance, it reported that the United Kingdom (UK) construction industry has faced decreasing performance due to issues of declining productivity and a lack of investments for research and training (Egan, 1998). Similarly, such problems have also surfaced in Singapore, where the local industry is reported to be extremely fragmented with numerous small firms lacking in capacity (Ministry of Trade and Industry, 2010). This runs contrary to the government's vision for Singapore's construction industry to 'develop into one which will encompass all aspect of the construction value chain, from design to maintenance' through the process of greater change and improvement (Ofori & Cuervo, 2005). As such, this paper will look into the underlying motivations for the prevalent use of subcontracting practices and the extent to which it has affected the local industry, such as declining productivity in the local industry.

For the purpose of this research study, productivity is defined as 'an attitude of the mind that strives to do things systematically, economically and with great attention to details so that what results is the best' (Low, 2015).

2.1.2 *Labour-Intensive Industry*

An important feature of the construction industry is that it is extremely labour intensive. It includes unskilled, semi-skilled, and skilled labour and staffs of technical and managerial positions. The unskilled and semi-skilled workers easily form the majority of the workforce that possesses little to no qualifications. The skilled labourers, on the other hand, are mainly tradesmen and onsite managers who have comprehensive understanding and are experienced in their specialised fields (Kirubaashini, 2015). Technical and management positions are typically filled by staff that have higher educational qualifications and are trained in areas such as drafting, administration, and command and control of the entire construction processes, as these positions are more technical and specialised. Examples of such

Table 2.1 Construction labour productivity (at 2010 market prices)

Year	Construction (%)
1996	−5.0
1997	−6.3
1998	−5.3
1999	−4.3
2000	0.0
2001	2.6
2002	−2.4
2003	2.4
2004	0.0
2005	0.5
2006	0.0
2007	7.0
2008	2.3
2009	8.1
2010	4.0
2011	2.0
2012	2.7
2013	−3.0
2014	−1.8
2015	0.5

Source Singapore Department of Statistics

areas include quantity surveying, architecture, civil engineering, etc. (Kirubaashini, 2015).

As a labour-intensive industry, there is a need to reduce the heavy reliance on manpower and strive to improve productivity in the industry. However, as shown in Table 2.1, there has been a decline in construction labour productivity within the past few years. As observed, from 2011 to 2015, there was a downward trend from 2.0 to 0.5% and at 2013, to just −3.0%, which indicates declining productivity levels. This is a cause for concern despite the government's efforts to improve productivity levels via initiatives such as the CPCF since 2010.

2.1.3 *Highly Susceptible to External Forces*

Generally, construction industries in almost all countries are also very susceptible to external forces, such as the volatility of the economy. For instance, the Latham Report mentioned that the UK's construction industry suffered a 39% decline in output in 1993 due to economic downturn (Latham, 1994). In response to the issue of economic downturn, the Latham and Egan Reports made recommendations for firms to engage in partnerships and alliances as an alternate form of project delivery. As defined by

Scott and Scott (2001), partnership is a relationship between two or more companies or organisations that is formed with the express intent of improving performance in the delivery of projects. This also entails the need for firms to collaborate more and share their resources to achieve a common outcome of project completion.

2.2 Practices Adopted in the Construction Industry

Even at present, construction industries worldwide have been practising outsourcing due to the popularity of specialisation. Eccles (1981) suggested that one common feature of the construction industry is the practice of subcontracting parts of a project by the main contractor to a specialised trade contractor (usually a subcontractor). Industry specialisation is an important determinant of subcontracting. With subcontracting, the number of players in a construction network or project will further increase, given that there are already many people involved in the construction project.

As such, this has resulted in a large number of small firms that employ fewer people due to their specialised nature. For instance, in the UK, there are an estimated 163,000 registered construction firms of which a large number employ less than eight employees (Orange, Burke, & Boam, 2000). Hence, subcontracting has been common in the industry due to the existence of numerous small firms. The same situation has also surfaced in Singapore, where Table 4 shows data extracted from the BCA, indicating that there is also a large proportion of small firms operating in Singapore's construction industry.

However, extensive subcontracting has reduced the continuity of working relationships, rendering such a segmented approach of project delivery and management inefficient, where it is blamed for the declining productivity levels in various countries (Cox & Townsend, 1998; Egan, 1998; Latham, 1994). Therefore, this suggests the need to overcome the problems of discontinuity and inefficiency that are the effects of fragmentation. Therefore, the coordination of specialised and differentiated tasks at the site level leads to the need for intense interaction between the players (Shirazi, Langford, & Rowlinson, 1996) in order to thwart the ill-effects of fragmentation.

2.2.1 Subcontracting Practices

The act of subcontracting is defined in the *Webster's Online Dictionary* (2006) as the outsourcing of a business to an individual or a business by a main contractor, for particular tasks of a whole project. Using subcontractors to complete portions of the overall work usually helps in bringing down outlays as well as in alleviating or avoiding project risks by transferring them to the third-party subcontractor. As these subcontractors are normally hired on a specialisation basis, the quality is assured for the main contractor compared to if the main contractor were to do the tasks in-house.

2.2.1.1 Types of Subcontracting

There are two main types of subcontracting practices commonly adopted by subcontractors, and they are described as follows (Watanabe, 1971):

- Specialisation-oriented subcontracting occurs when parent firms and their subcontractors are engaged in dissimilar but complementary activities. They might be working towards final products of a similar form but through different processes or might be making different final products in order to expand the parent firm's range.
- Capacity-oriented subcontracting occurs when both the parent firm and subcontractor are engaged in a similar trade of work but are mutually competitive by nature, in which they would become competitors once demand decreases.

2.2.1.2 Reasons for the Adoption of Subcontracting Practices

The reasons for the adoption of the two main types of subcontracting practices are described as follows (Watanabe, 1971):

- Specialisation-oriented subcontracting mainly occurs due to the lack of specialised equipment or skills. This is to take advantage of a subcontractor's specialised technology and skillsets. In construction, it is almost impossible for any company to cover all aspects of a construction project due to the need for diverse skillsets and technology.
- Capacity-oriented subcontracting is used to provide a buffer against business fluctuations to avoid the need to maintain redundant capacity during low business cycles but still be able to meet peak demand without having to provide for excessive overhead during non-peak times. It is usually adopted by larger parent firms to meet peak demand.

2.2.2 *The Multi-Layered Subcontracting System (MLSS)*

The MLSS is a hierarchy-based subcontracting system that is commonly adopted in construction industries worldwide, but it is more prevalent in Singapore. This happens when a subcontractor is involved in multiple projects and further subcontracts work to other subcontractors (Ofori & Lim, 2009). It starts with a single layer of subcontracting, where one subcontractor subcontracts work out to another subcontractor. Following this, multiple layers could subsequently exist, depending on the workload and frequency of subcontracting. This in turn creates a pyramid of subcontracting relationships, where layers start to flare downstream, and the eventual product is a form of MLSS. In the meantime, the parent subcontractor maintains a profit on the work, while still being responsible for the quality of

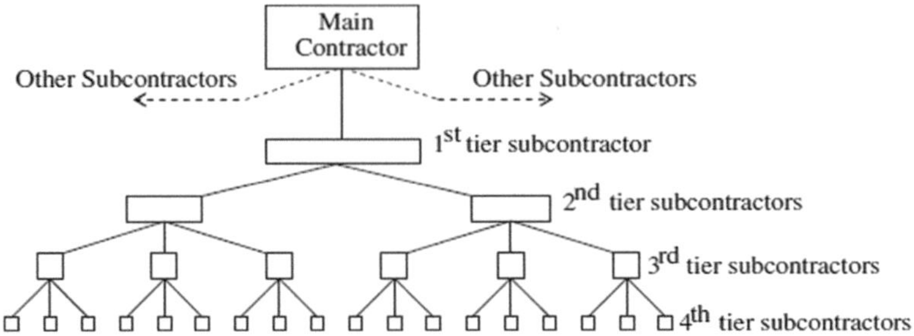

Fig. 2.1 Pyramid of subcontractors. *Source* Yik and Lai 2008

workmanship to the main contractor (Ofori & Lim, 2009). A graphical display of MLSS is shown in Fig. 2.1.

In addition, such an arrangement is established to reduce transaction costs, where there is little or no need to form detailed contracts and to source for other subcontractors when there are new projects (Yik & Lai, 2008). It is important to note that such implicit business relationships are commonly practised, especially in Singapore. They are usually for the long term, with both parties mutually agreeing to collaborate together, but they are not contractually bound.

2.2.2.1 Peculiarities of MLSS

MLSS and Specialist Subcontracting

From the project commencement till its completion, the workload of the employees of the subcontractor varies in intensity level according to their trades. For example, different trades, such as equipment installation, testing, and commissioning, all have varying periods of work intensity. This happens when these core pools of employed workers from the various trades have lots of idle time when a subcontractor undertakes all work in-house, where such idling manpower is a source of inefficiency. In order to overcome such resource inefficiency, the adoption of subcontracting out further projects in various work packages is carried out.

To reduce cost, most primary subcontractors or, at times, even main contractors engage in secondary subcontracting for discrete trades of work by hiring other subcontractors to do their work (Yik, Lai, Chan, & Yiu, 2006a, b, c). Hence, such a move is typical of capacity-oriented subcontracting (Yik & Lai, 2008), where this is a form of MLSS, as such subcontracting at the lower tiers is principally adopted for effective organisation of manpower in order to cope with variable and fragmented work demands during various business cycles. However, the downsides of such capacity-oriented subcontracting are that it fails to capitalise and harness the true potential of the specialisation of lower tier subcontractors, reduces the value added per subcontracting tier, and, most disastrously, significantly reduces the profit margins of subcontractors that are in the lower tiers of the MLSS.

Despite such downsides, subcontractors in the lower rungs still continue to do multiple projects simultaneously and usually work under multiple employers on a job basis (Yik et al., 2006a, b, c). By working for a number of parent firms, subcontractors are able to maintain a more consistent workforce level, thereby economising on transaction costs, which would result in high turnover of workers in response to demand volatility (An & Watanabe, 2011). Such an occurrence could be explained due to the fact that, in most construction trades, physical capital investment requirements are quite low and not highly idiosyncratic, while human capital investments are more important, and these specialised trade subcontractors can obtain production economies. As such, it also essentially means that a constant workforce results in the keeping of a core pool of workers (long-term staff) and less reliance on peripheral workers (temporary staff).

Reduction in Profit Margins Down the MLSS Hierarchical Tiers

As defined earlier, the definition of MLSS encompasses the parent subcontractor maintaining a profit on the work that is subcontracted out to another secondary subcontractor down the hierarchy, while still being responsible for the quality of workmanship to the main contractor. As such, every succession down the MLSS hierarchical tier of subcontractors typically deducts a compounded percentage of profit kept by the parent subcontracting firm. The profit to be made by the secondary subcontractors within the MLSS hierarchy is based on the delivery of a prescribed amount of work by the parent subcontracting firm. As a result, the amount of profit to be made by subcontractors within the lower rungs of the MLSS for succeeding work packages would be gradually lowered as the work is further subcontracted downstream of the MLSS pyramid, eventually being reduced to very low piece-rates (Yik & Lai, 2008). Hence, subcontractors at the lowest rung will be constantly strained for resources, both monetary and manpower.

Lack of Value-Added Subcontracting in MLSS

The first sign of distress in an MLSS framework is evident whenever the main contractor does not conduct supervisory checks over the work of subcontractors, which is where construction quality tends to suffer. This commonly occurs when the profit margins of the lower tier subcontractors are progressively being reduced down the MLSS hierarchy. As such, the subcontractors at the lower tiers may be enticed to take chances by cutting corners to cut costs at the expense of quality (Construction Industry Review Committee, 2001). This practice is essentially the main culprit for the decreases in value-add down the project delivery chain and is possibly a huge contributor to the reduction in productivity. Hence, it is imperative for contractors to pay attention to the practice of broker-type subcontracting where multiple non-value adding layers complicate communication and progressively reduce the profit margin of the lower tier subcontractors, which is possibly the root cause of the lack of added value in an MLSS framework.

Labour-Only Subcontracting Practice at the Lowest Tier of MLSS

For many local construction firms, it is conventional practice to engage in subcontracting, and most subcontract a great extent of their work on a labour-only basis (Ofori & Debrah, 1998). Ofori and Debrah (1998) argued that the cause of labour subcontracting is in response to (1) the predominance of labour-intensive construction techniques adopted in the industry, (2) the acute labour shortage in Singapore, (3) the industry's poor image and hence its inability to attract local personnel, (4) the uncertain work environment, and (5) acceptance of and familiarity with a system that has become entrenched owing to a long history (Ofori & Debrah, 1998). Hence, the *kepala* (head) system that is commonly known to many contractors has been researched extensively and is blamed for the low construction productivity in Singapore (Ofori & Debrah, 1998).

2.2.2.2 Controversies Associated with MLSS

Up till today, most international literature has focused on the phenomenon of MLSS, its associated difficulties in project management (Tam, Shen, & Kong, 2011), and the general wellbeing of construction workers under the MLSS (Yoon & Kang, 2003).

In Hong Kong, studies have discovered that inappropriate work practices in an MLSS have caused instances of poor quality delivery, stretched communication chains that contributed to poor time management, increased instances of abortive and remedial work from these inappropriate practices, amplified budget overruns, and failed communication, leading to poor coordination of processes (Tam et al., 2011). Other researchers in Hong Kong have demonstrated how MLSS was considered one of the contributing factors for poor safety performance, as subcontractors are occasionally small and may lack safety commitments due to limited resources and time (Wong & So, 2004). Further research has even suggested that legislation ought to be enacted to restrict subcontracting practices in order to improve safety on site (Linehan, 2000; Tong, 2000).

Elsewhere in the UK, studies have put blame on MLSS for the high accident rates due to the loss of managerial control resulting from multiple tiers of subcontracting, fragmenting the safety responsibility of main contractors (Harvey, 2003). Moreover, in Korea, the oligopolistic market structure and MLSS were blamed for low quality of construction products and illegal construction practices, such as bribery, and have resulted in a labour force with little training and low productivity (Yoon & Kang, 2003).

Based on the review of the literature, the system of multi-layered subcontracting has had negative implications for construction industries in many countries. Therefore, it is imperative that ample attention is devoted to researching whether the MLSS has indeed had a similar negative effect on the local construction industry. If so, how best can the industry cooperate and handle such negative implications for the better?

2.2.2.3 Successful Implementation of MLSS in Japan

Despite the seemingly negative facets of MLSS that have been discussed above, MLSS could also be successful and beneficial for the construction industry if the proper measures are taken. The Japanese construction industry has many multi-layered business relationships. The structure of the Japanese construction industry is one where main contractors organise, finance, and manage the project. Specialist subcontractors are hired on a project basis rather than employed in-house, as a way to increase labour flexibility. Japanese main contractors have a tendency to work with the same nucleus of subcontractors and vendors and, apart from providing work, also provide training, skills upgrading, recommendations for new tools and processes, and even basic management expertise (Lamert, Funato, & Poor, 1996). As these parties continue to work together from project to project, the relational contracting mode appears.

The concept of affiliated companies is a non-contractual cooperation between different layers of subcontractors, which results in risk sharing and confidence, and it is a significant contributor to the positive performance of MLSS (An & Watanabe, 2011). The MLSS is heralded as a great contributor to the success of the Japanese construction industry in that it facilitates the increasing specialisation of the subcontracting market with plentiful small and medium-sized specialist subcontractors, who make incessant efforts on improving technical and engineering issues and on finding ways to accord skills to their workers. This relationship is also built on trust, as Japanese contractors generally establish non-contractual cooperative relationships down the construction supply chain (An & Watanabe, 2011). Therefore, it is with hope that in the near future, the local industry can take a step in this direction to improve the MLSS structure.

References

An, T., & Watanabe, T. (2011). Towards the sustainable improvement in construction labour's employment in China: Cultivating subcontractor's role. In C. Egbu & E. C. W. Lou (Eds.), *Proceedings 27th Annual ARCOM Conference* (pp. 663–672), 5–7 September 2011, Bristol, UK, Association of Researchers in Construction Management.

Construction 21 Steering Committee. (1999). *Construction 21: Re-inventing construction.* Ministry of Manpower and Ministry of National Development, Singapore.

Construction Industry Review Committee. (2001). *Chapter 4: Fostering a Quality Culture* (pp. 34–52). Construct for Excellence, Construction Industry Review Committee, Hong Kong.

Cox, A., & Townsend, M. (1998). *Strategic procurement in construction: Towards better practice in the management of construction supply chains*. London: Thomas Telford.

Eccles, R. (1981). The quasifirm in the construction industry. *Journal of Economic Behaviour and Organization, 2*(4), 335–357.

Egan, J. (1998). *Rethinking construction: The report of the construction task force*. London: HMSO.

Foo, C. L., & Ho, C. L. (2001). *Critical success factors of supply chain management for theconstruction industry in Singapore*. Unpublished Master's Thesis, BCA Centre for AdvancedConstruction Studies, Nanyang Technological University, Singapore.

Harvey, M. (2003). The United Kingdom: Privatization, fragmentation, and inflexible flexibilization in the UK construction industry. In G. Bosch & P. Philips, *Building Chaos: An international comparison of deregulation in the construction industry* (pp. 188–209). London: Routledge.

Kirubaashini, D. (2015). *Role of E-HRM in achieving HRM performance in construction firms in Singapore* [Unpublished BSc. (Building) study]. National University of Singapore, Singapore.

Lamert, B. H., Funato, K., & Poor, A. (1996). *The construction industry in Japan and its subcontracting relationships.* (EIJS WP no. 21): Working Paper for the European Institute of Japanese Studies, Stockholm School of Economics.

Latham, M. (1994). Constructing the team: Final report: *Joint review of procurement and contractual arrangements in the United Kingdom construction industry*. London: HMSO.

Linehan, A. (2000). *Subcontracting in the construction industry*. A speech presented in Safety and Health Conference. Hong Kong. Retrieved March 22–23, 2000.

Low, C. (2015). *Productivity Concepts and Techniques* (Unpublished Lecture notes). Building and Construction Authority Academy, Singapore.

Ministry of Trade and Industry. (2010). *Construction Working Group Report* (pp. 94). Singapore: Economic Review Committee, MTI.

Morledge, R., Knight, A., & Grada, M. (2009). The concept and development of supply chain management in the UK construction industry. In S. Pryke (Ed.), *Construction supply chain management: Concepts and case studies* (pp. 23–39). London: Wiley.

Ofori, G., & Cuervo, J. (2005). *Consortia for export of construction services in Singapore. Global perspectives on management and economics in the AEC sector* (pp. 42–56). Finland: Technical Research Centre of Finland.

Ofori, G., & Debrah, Y. A. (1998). Flexible management of worker: Review of employment and practices in the construction industry in Singapore. *Construction Management and Economics, 16*(4), 397–408.

Ofori, G., & Lim, K. H. (2009). Multi-layered subcontracting system in Singapore's construction industry. In K. Kobayashi, A. R. Kharuddin, G. Ofori, & S. Ogunlana (Eds.), *Joint ventures in construction* (pp. 216–226). Great Britain: Thomas Telford Limited.

Orange, G., Burke, A., & Boam, J. (2000). *Organisational learning in the UK construction industry: A knowledge management approach*. Paper presented at ECIS, Vienna, Austria, July 5–7, 2000.

Schwegler, B. R., Fischer, M. A., O'Connell, M. J., Hanninen, R., & Laitinen, J. (2001). *Near-, medium-, and long-term benefits if information technology in construction. Center for integrated facility engineering (CIFE)* (Working paper #65), Stanford University, Stanford, CA.

Scott, B., & Scott, R. (2001). What is partnering—Why adopt alliancing. In *Partnering in Europe: Incentive based alliancing for projects*. London: Thomas Telford.

Shirazi, B., Langford, D., & Rowlinson, S. (1996). Organizational structures in the construction industry. *Journal of Construction Management and Economics, 14*(3), 199–212.

Tam, V. W., Shen, L., & Kong, J. S. (2011). Impacts of multi-layer chain subcontracting on project management performance. *International Journal of Project Management, 29*(1), 108–116.

Tong, Y. (2000). *Scrapping of rotten subcontracting system urged*. An article of interview by South China Morning Post.

Watanabe, S. (1971). Subcontracting, industrialisation and employment creation. *International Labour Review, 104*(1/2), 51–76.

Wong, F. K., & So, L. S. (2004). Multi-layers subcontracting practice in the Hong Kong construction industry. In S. Rowlinson, *Construction Safety Management System* (pp. 137–148). London: Taylor & Francis Group.

Yik, F. W., & Lai, J. H. (2008). Multilayer subcontracting of specialist works in buildings in Hong Kong. *International Journal of Project Management, 26*(4), 399–407.

Yik, F., Lai, J., Chan, K. T., & Yiu, E. (2006a). *Best practices in managing specialist subcontracting performance*. Hong Kong: The Hong Kong Polytechnic University. Retrieved May 13, 2014 from http://repository.lib.polyu.edu.hk/jspui/bitstream/10397/2347/1/Full_Report_on_Best_Practices_in_Managing_Specialist_Subcontracting_Performance.pdfFresearch.

Yik, F., Lai, J., Chan, K., & Yiu, E. (2006b). *Best practices in managing specialist subcontracting practices* (Final Research Project Report). Hong Kong: Construction Industry Institute.

Yik, F. W. H., Lai, J. H. K., Chan, K. T., & Yiu, E. C. Y. (2006). Problems with specialist subcontracting in the construction industry. *Building Services Engineering Research & Technology, 22*(1), 183–193.

Yoon, J. H., & Kang, B.-G. (2003). Republic of Korea—*The drivers for change in the Korean construction industry: Regulation and deregulation*. In G. Bosch, & P. Philips (Eds.), *Building Chaos: An international comparison of deregulation in the construction industry* (pp. 210–233). London: Routledge.

Chapter 3
Case of Singapore's Construction Industry

Abstract This chapter assesses the level of fragmentation in the Singapore construction industry using statistical data from Singapore's Building and Construction Authority that spans from 1997 to 2015. The statistical analysis observes a large presence of small and medium-sized firms that have increased over the years. This suggests that fragmentation, in terms of firm size, still exists within the industry. In addition, subcontracting practices are still prevalent within the industry.

As discussed in earlier sections, the construction industry in general is fraught with low productivity, ineffective resource management, uncontrolled budgets, and defective work, and this is the same predicament for Singapore as well (Lim & Low, 1992). This is despite the many management innovations that have been implemented, as persistent problems still exist within the industry. The acknowledgement of these persistent problems has prompted the Singapore government to initiate the formation of the Construction 21 Steering Committee by the Ministry of Manpower and Ministry of National Development (Construction 21 Steering Committee, 1999). From there, more sustainable improvements in construction management are further highlighted. In addition, following the government's acknowledgement of the strategic importance of the industry to the country's economic progress, the government has continuously been embarking on efforts in transforming the construction industry in Singapore. Currently, the local industry has been divided into three categories under the Singapore Standard Industrial Classification (SSIC), namely, construction of buildings, civil engineering, and specialised construction activities (Department of Statistics, 2010).

L. Sui Pheng and S. Jia Zheng, *Construction Productivity in the Multilayer Subcontracting System*, Management in the Built Environment,
https://doi.org/10.1007/978-981-13-1831-3_3

3.1 Issues Surrounding Singapore's Construction Industry

The Construction 21 report was produced by the Singapore government in an attempt to transform perceptions of the industry from a dirty, demanding, and dangerous industry to a professional, productive, and progressive industry (Ofori, 2003). This report also surfaced several problems within the local industry that were similar to those reported in international reports of other countries' construction industries. One such report is the Egan Report that depicts the scope for improving the quality and efficiency of the UK construction industry.

A key focus of this study is that the local construction industry is considered fragmented. According to a survey done by the Building and Construction Industry (BCA, 2016), there are about 13,909 contracting firms in the industry (Ministry of Trade and Industry, 2010). This problem was also raised in the Egan Report which reported the highly fragmented nature of the construction industry in UK. In response to the highly fragmented nature of the industry, the Egan Report suggested that this may be both a strength and a weakness. The strength included the agility of smaller firms to be highly flexible in their ability to deal with a varied workload in different economic phases. The downside, however, was the fact that the presence of numerous small firms encouraged subcontracting. As such, the prevalence of such subcontracting practices has come to the point at which it has significantly reduced the continuity of contractual relationships between firms that was essential to improving efficiency. This subsequently led to a decline in construction productivity.

3.1.1 Current Subcontracting Framework in Singapore

The high extent of subcontracting in Singapore's construction industry is indeed a norm, where around 60–70% of construction work is subcontracted (Loh & Ofori, 2000). To reiterate, it is a common practice by main contractors to engage subcontractors for their specialised skills to supplement labour and/or material to perform construction work. Figure 3.1 shows a typical construction project structure in Singapore.

3.1.1.1 Subcontracting in the Local Context

In Singapore, main contractors usually hire subcontractors for materials only, labour only, or both. Labour-only subcontracting is prevalent in Singapore's construction industry, where subcontractors provide the labour component of each trade to the main contractor, based on a project-to-project basis that is of a temporary arrangement when required. In terms of obligations, the main contractor will be responsible for the provision of machinery, equipment, material, and supervision

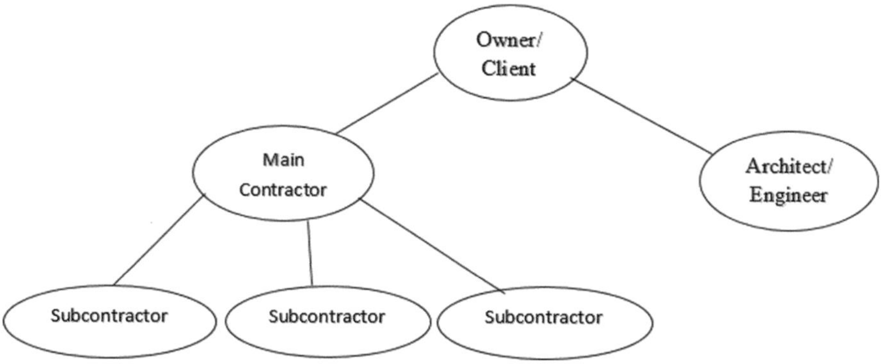

Fig. 3.1 Construction project structure. *Source* Lim, 2005

as well as office management on site, whereas the subcontractors will be in charge of recruitment, control, salary, and the provision of the necessary small tools and trade specialisation skills when called upon (Debrah & Ofori, 1997). As such, a channel is created between contractors who can provide work opportunities and workers who are actively seeking jobs (Wells, 2006).

Furthermore, when subcontractors have too many projects on hand, there is a tendency of subletting the work to other or smaller subcontractors. This helps them to cope more flexibly with the variable workload and may even allow them to earn extra profit. Moreover, the work may be further sublet, and this gives rise to multi-layered subcontracting (Ofori & Debrah, 1998). As such, it is worthwhile to note that Singapore's subcontracting framework closely mirrors that of MLSS, due to the many similar characteristics shared by both.

However, despite the fact that local research studies have noted that the Singapore construction industry considers MLSS a key cause of many of the problems (Ofori & Lim, 2009), this system remains largely unregulated in Singapore. With that, the Singapore construction industry is plagued by many problems from the MLSS, to the point that it has been labelled the culprit of numerous issues, such as substandard construction quality assurance, low productivity because of fatigue of workers being shuttled between sites, shortage of incentive for subcontractors to increase the qualifications of their workers, subversion of the accountability of subcontractors due to the lack of privity of contract between the main contractor and the lower tier subcontractors, and gaps in standard contract clauses that are sometimes overlooked due to project time constraints (Ofori & Lim, 2009).

3.1.1.2 Types of Subcontractors in the Local Context

The three main categories of subcontractors in the construction sector are classified in Table 3.1.

Table 3.1 Categories of subcontractors

Classification of subcontractor categories	Characteristics
Trade subcontractors	Perform general trade services, such as brickwork, formwork, painting, and finishing work
Specialist subcontractors	Undertake specialist services, such as air-conditioning, electrical, elevator, and plumbing
Labour-only subcontractors	Offer labour-only services. Main contractor will provide the materials and does supervision and management of the work

Source Hinze and Tracy, 1994

In terms of contractual view, there are also three categories of subcontractors (Mbachu, 2008). The first is termed 'domestic subcontractors'. They are engaged by the main contractor to perform parts of their work. Commonly, domestic subcontractors are employed for trade services. The second category is 'selected subcontractors', where they are solicited from a recommended list of potential subcontractors in the tender documents. Lastly, there are 'nominated subcontractors'. Clients will select and nominate them to the main contractor to carry out specified tasks in the contract. Hence, nominated subcontractors usually undertake specialist services. Apart from nominated subcontractors, the lowest bidder (subcontractor) is normally chosen. However, for the purpose of this study, this paper will combine the three classifications of subcontractors into one generic form of subcontractors to aid in easier analysis and comparison of other data.

3.2 Statistical Analysis of Singapore's Construction Industry

A detailed analysis is conducted on data acquired from BCA's "Survey of the Construction Industry", spanning from 1996 to 2015 and specifically pertaining to the related issues of fragmentation and subcontracting in the industry. The analysis seeks to highlight the changes in the number of construction firms, both large and small, and their respective construction turnover and employment size among large and small firms. These changes will serve to establish the state of Singapore's construction industry and validate numerous literature studies that show the MLSS is indeed prevalent here. Thereafter, deductions can be made as to whether the widespread practice of subcontracting has an effect on construction productivity.

Analysis of Table 3.2

As observed in Table 3.2, there has been a 47% increase in the number of enterprises from 1996 to 2013. Concurrently employment has increased by as much as 91% within the same period. This points to a booming industry where the presence of multiple projects in the market has increased turnover from approximately SGD

Table 3.2 Overall key indicators of construction industry

Year	No. of enterprises	No. of employment	Construction turnover ($,000)
1996	11,415	229,595	35,630,438
1997	13,496	278,725	40,911,902
1998	13,954	266,196	39,877,196
1999	13,909	250,599	36,665,562
2000	13,484	256,331	35,354,933
2005	12,979	215,371	27,867,645
2010	14,810	355,515	51,344,721
2011	15,434	376,443	54,584,774
2012	15,673	401,527	61,126,411
2013	16,823	438,927	68,021,906

$35 billion in 1996 to SGD $68 billion in 2013. Overall, the various trends have shown a positive indication that the construction industry is expanding within the last two decades.

Analysis of Table 3.3

A closer look at Table 3.3 more specifically illustrates the industry trends within the past decade from 2005 to 2013. For ease of readability and relevance, the data have been categorised into three horizontal sections, each focusing on small, medium, and large firms, respectively, from data that have been extracted from BCA's 'Survey of the Construction Industry' booklets over the years. For the purpose of this research, the classification standard on the varying firm size is based on Gartner, Inc., the world's leading information technology research and advisory company. It defines businesses using the commonly used attribute of the number of employees. Small businesses are usually defined as organisations with fewer than 100 employees, while midsize enterprises are those organisations with 100 to 999 employees. However, as BCA's data only reflects up to 500 employees and above, firms with more than 500 employees are classified as large, followed by medium-sized firms with between 100 and 499 employees, and lastly, those employing less than 100 employees are considered small.

More notably, there was an approximate 28% increase in the number of small firms from 2005 to 2013, where concurrently, there was also an increase in both the number of people employed and the construction turnover. With that, operating profit has dramatically increased from approximately SGD $33 million to SGD $898 million within a decade. This could be an indication of the relatively low barriers to entry into the industry, with the availability of construction projects in a booming market, which has helped spur the growth of small construction firms. This is also similar to both the medium and large firms, as their operating profits have largely increased over the past decade.

Therefore, it can be concluded that the large presence of small and medium firms that have increased over the years is proof that fragmentation, in terms of firm size, still firmly exists within the industry. In addition, it can be inferred that

Table 3.3 Overall key indicators of construction industry

Year	2005	2010	2011	2012	2013
Small firms (<5 to 99 employees)					
No. of enterprises	12,617	14,099	14,558	14,883	16,139
No. of employment	138,594	189,605	198,213	212,722	247,873
Construction turnover ($,000)	15,630,791	22,461,946	24,601,304	27,133,958	28,342,642
Operating expenditure ($,000)	15,597,411	21,747,215	23,824,814	26,494,063	27,443,850
Operating profit ($,000)	33,380	714,731	776,490	639,895	898,792
Medium firms (100 To 499 Employees)					
No. of enterprises	350	663	831	721	604
No. of employment	68,168	128,152	140,555	128,954	118,997
Construction turnover ($,000)	11,308,764	18,635,112	19,545,647	19,706,853	21,615,150
Operating expenditure ($,000)	11,222,265	17,410,625	18,491,743	18,864,937	21,307,245
Operating profit ($,000)	86,499	1,224,487	1,053,904	841,916	307,905
Large firms (>500 employees)					
No. of enterprises	12	49	45	69	81
No. of employment	8608	37,756	37,677	59,851	72,058
Construction turnover ($,000)	928,090	10,247,662	10,437,820	14,285,599	18,064,115
Operating expenditure ($,000)	1,002,581	9,644,861	9,791,549	13,642,701	17,480,960
Operating profit ($,000)	−74,491	602,801	646,271	642,898	583,155

subcontracting practices are still prevalent within the industry. As such, the next chapter will attempt to explain the reasons for this through the literature review on the rampant practice of subcontracting in Singapore. As this is a qualitative research study, efforts to determine the effect on construction productivity due to subcontracting will be in the form of detailed interviews with both main contractors and subcontractors. Responses from the interviews will be analysed via word mining tools to gauge the effect on construction productivity.

References

Construction 21 Steering Committee. (1999). *Construction 21: Re-inventing construction.* Ministry of Manpower and Ministry of National Development, Singapore.

Debrah, Y. A., & Ofori, G. (1997). Flexibility, labour subcontracting and HRM in the construction industry in Singapore: Can the system be refined? *International Journal of Human Resource Management, 8*(5), 690–709.

Hinze, J., & Tracy, A. (1994). The contractor-subcontractor relationship: The subcontractors views. *Journal of Construction Engineering & Management, 2*(1), 274–287.

Lim, K. H. (2005). *Causal management of subcontractors in waste minimization* (Unpublished doctoral study, School of Design and Environment). National University of Singapore, Singapore.

Lim, L. Y., & Low, S. P. (1992). *Just in time productivity for construction.* Singapore: SNP publishers.

Loh, W. H., & Ofori, G. (2000). Effect of registration on performance of construction subcontractors in Singapore. *Engineering, Construction and Architectural Management, 7*(1), 29–40.

Mbachu, J. (2008). Conceptual framework for the assessment of subcontractors eligibility and performance in the construction industry. *Construction Management & Economics, 26,* 471–484.

Ministry of Trade and Industry. (2010). *Construction working group report* (pp. 94). Singapore: Economic Review Committee, MTI.

Ofori, G. (2003). Preparing Singapore's construction industry for the knowledge-based economy: Practices, procedures and performance. *Construction Management and Economics, 21*(2), 113–125. https://doi.org/10.1080/0144619032000079680.

Ofori, G., & Debrah, Y. A. (1998). Flexible management of worker: Review of employment and practices in the construction industry in Singapore. *Construction Management and Economics, 16*(4), 397–408.

Ofori, G., & Lim, K. H. (2009). Multi-layered subcontracting system in Singapore's construction industry. In K. Kobayashi, A. R. Kharuddin, G. Ofori, & S. Ogunlana (Eds.), *Joint ventures in construction* (pp. 216–226). Great Britain: Thomas Telford Limited.

Wells, J. (2006). Labour subcontracting in the construction industries of developing countries: An assessment from two perspectives. *Journal of Construction in Developing Countries (JCDC), 11*(1), 17–36.

Chapter 4
Theoretical Review for Singapore's Subcontracting Practices

Abstract This chapter draws upon ideas and theories of past studies for the reasons why firms still adopt subcontracting practices in a fragmented industry. This forms the basis of understanding how subcontracting has an effect on industry-wide construction productivity. The combinative explanatory framework presented in this chapter depicts a single construction firm having to manage its own resources and competencies in order to achieve competitive advantage.

This chapter covers past literature to provide a comprehensive theoretical explanation of the current circumstances of Singapore's construction industry and adopted practices and outlines the effects of such adopted practices in Singapore.

4.1 Resource-Based View Approach

As discussed in earlier sections, fragmentation in Singapore's construction industry is prevalent, which leads to the rampant practice of subcontracting. The underlying reasons for these occurrences could be partially explained by the renowned resource-based view (RBV) approach. According to Porter (1991), the RBV is part of a key theory involving competitive advantage, and this literature is more prominent and more extensively adopted and, as such, deserves more in-depth study. Concurring with his view, Kraaijenbrink, Spender, and Groen (2009) mentioned that the RBV theory has become one of the most influential and cited theories in the history of management theorising. The prominence of the RBV theory has its origins in the realm of competitive advantage, where the presence of valuable resources or competences that firms possess are often intangible assets, such as skills, reputation, etc. These resources are seen as relatively immobile, which constitute strengths that should be nurtured (Porter, 1991).

Basically, in the RBV model, resources have a major role in helping companies achieve higher organisational performance (Barney, 1991; Rothaermel, 2012). There are two types of resources: tangible and intangible. Tangible assets are

L. Sui Pheng and S. Jia Zheng, *Construction Productivity in the Multilayer Subcontracting System*, Management in the Built Environment,
https://doi.org/10.1007/978-981-13-1831-3_4

physical resources that can easily be bought in the market; thus, they confer little advantage to the companies in the long run. Intangible assets are everything else that has no physical presence but can still be owned by the company. Brand reputation, trademarks, and intellectual property are all intangible assets. Unlike physical resources, brand reputation is built over a long time and is something that other companies cannot buy from the market. Intangible resources usually stay within a company and are the main sources of sustainable competitive advantage.

In addition, advocates of RBV have proposed that firms look inside their own respective companies to find the sources of competitive advantage instead of looking at the competitive environment and that it is much more feasible to exploit external opportunities using existing resources in a new way rather than trying to acquire new skills for each different opportunity (Barney, 1991; Rothaermel, 2012).

4.1.1 RBV Framework and Its Assumptions

The RBV theory, like any other theory, operates based on a set of key assumptions. An illustration of the RBV framework is also shown below for a better understanding of this theory.

4.1.1.1 Overall RBV Framework

The RBV framework in Fig. 4.1 depicts the various factors at play that enable firms to achieve competitive advantage and seeks to explain each of its components that contributes to a firm's competitive advantage (Rothaermel, 2012).

4.1.1.2 VRIO Framework

Although having heterogeneous and immobile resources is critical in achieving competitive advantage, it is not enough if the firm wants to sustain it. Barney (1991) has identified a valuable, rare, inimitable, and non-substitutable (VRIN) framework that examines whether resources are valuable, rare, costly to imitate, and non-substitutable. The framework was later improved from VRIN to VRIO by determining whether the firm is organised to be able to exploit these resources (Rothaermel, 2012). In determining whether a firm is able to attain sustained competitive advantage (SCA), the following questions are asked:

- **Question of Value**. Resources are valuable if they help organisations to increase the value offered to the customers. This is done by increasing differentiation or/and decreasing the costs of the production. The resources that cannot meet this condition lead to competitive disadvantage.

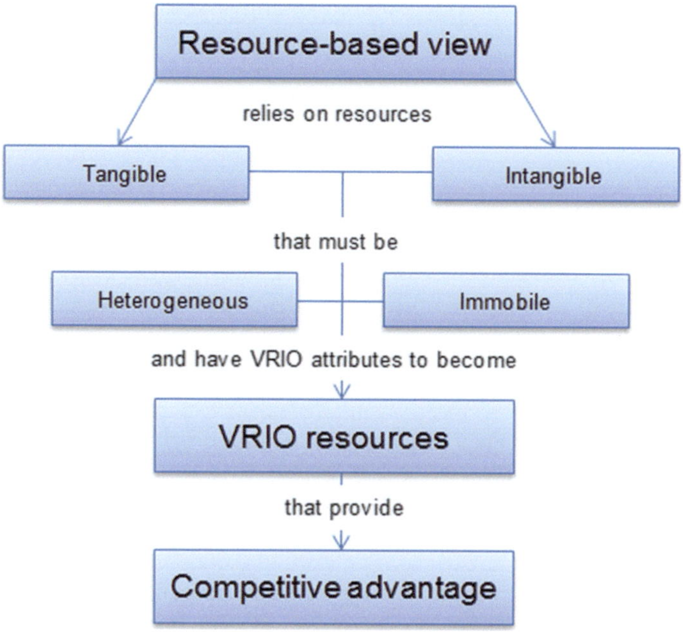

Fig. 4.1 Overall RBV framework diagram

- **Question of Rarity**. Resources that can only be acquired by one or a few companies are considered rare. When more than a few companies have the same resource or capability, it results in competitive parity.
- **Question of Imitability**. A company that has a valuable and rare resource can achieve at least temporary competitive advantage. However, the resource must also be costly to imitate or to substitute for a rival if a company wants to achieve SCA.
- **Question of Organisation**. The resource itself does not confer any advantage for a company if it is not organised to capture the value from them. Only the firm that is capable of exploiting the valuable, rare, and imitable resources can achieve SCA.

A firm will achieve SCA if it has the resources and capabilities to answer 'yes' to all questions that are posed in the framework, as shown in Fig. 4.2.

The central premise of the RBV approach is about sustaining a firm's competitive advantage, such that they can effectively outcompete their rivals within the industry. Peteraf and Barney have defined competitive advantage as:

> An enterprise has a Competitive Advantage if it is able to create more economic value than the marginal (break-even) competitor in its product market. […] The Economic Value created by an enterprise in the course of providing a good or service is the difference between the perceived benefits gained by the purchasers of the good and the economic cost to the enterprise (2003, p. 314).

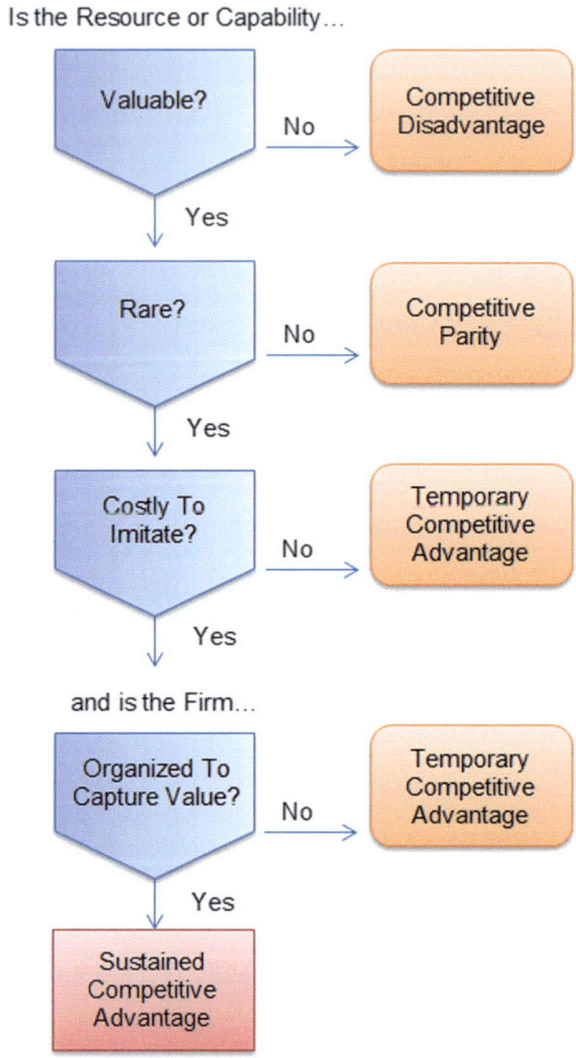

Fig. 4.2 VRIO framework

As such, for a firm to achieve a state of SCA, it must acquire and control VRIN resources and capabilities from within the firm and have the organisation in place that can absorb and apply them (Barney, 1991, 1994, 2002). This proposition is shared by several related analyses, such as core competences (Hamel & Prahalad, 1994), dynamic capabilities (Helfat & Peteraf, 2003; Teece, Pisano, & Shuen, 1997), and the knowledge-based view (Grant, 1996). The implicit focus of RBV is on the underpinnings of the successful diversification of a firm's resources, such as the adoption of subcontracting practices by firms in the local construction industry context to minimise risks (Porter, 1991). Therefore, it is also essential when diversifying to understand a firm's distinctive strengths, for instance, the trade specialisation of a particular firm.

4.1.1.3 Assumptions of the RBV Approach

The RBV approach adopts two assumptions in analysing sources of competitive advantage (c.f. Barney, 1991; Peteraf & Barney, 2003). First, this model assumes that firms within an industry (or within a strategic group) may be heterogeneous with respect to the bundle of resources that they control (Bridoux, 2004). It means that the skills, capabilities, and other resources that organisations possess differ from one company to another. If organisations would have the same amount and mix of resources, they could not employ different strategies to outcompete each other (Rothaermel, 2012). Second, it assumes that resource heterogeneity may persist over time because the resources used to implement firms' strategies are not perfectly mobile across firms (i.e., some of the resources cannot be traded in factor markets and are difficult to accumulate and imitate). Resource heterogeneity (or uniqueness) is considered a necessary condition for a resource bundle to contribute to a competitive advantage (Bridoux, 2004). The argument is 'If all firms in a market have the same stock of resources, no strategy is available to one firm that would not also be available to all other firms in the market' (Cool, Almeida, Costa, & Dierickx, 2002, p. 57).

4.1.2 Applicability of RBV Approach to Local Construction Industry

As firms in the construction industry strive to compete with each other, naturally their first instinct is to search for opportunities to exploit via adopting the most appropriate strategy. As mentioned, one of the key assumptions of RBV is the immobility of resources. Although this gives some form of competitive advantage for a firm over their rivals in areas such as brand equity, processes, and intellectual property, all these are just temporary. In order to outlast, outwit, and outcompete their competitors, the practice of subcontracting by most firms is deemed to provide a SCA over their rivals. This is usually a key strategy employed by construction firms here and abroad. Hence, most firms resort to subcontracting their work out to another party, or at times, to multiple parties. The challenge, however, would be to ensure smooth coordination of specialised and differentiated tasks at the site level, which will lead to the need for intense interaction between the players (Shirazi, Langford, & Rowlinson, 1996).

For instance, in Singapore's construction industry, all construction firms face constraints in their availability of resources in terms of labour, capital, and time. Key challenges in Singapore's construction industry that were highlighted by the Economic Review Committee in its construction working group report include an overcapacity of local construction resources, the upgrade capabilities of local firms to compete with established foreign firms, and the size and financial constraints faced by the construction sector when venturing overseas. Moreover, other issues include the presence of numerous small firms that do not encourage industry upgrading and economies of scale and that many contractors are still constrained by costs and time to initiate or adopt better techniques or invest in research and development.

Therefore, with respect to the RBV theory, construction firms must navigate such treacherous environments by employing a suitable strategy based on the availability of the firm's resources. For instance, a reputable local main contractor like Woh Hup Pte. Ltd. can employ its intangible assets like its strong brand reputation and its intellectual property in construction automation to promote its construction services in the industry to prospective clients, thereby enabling it to attain more projects than its rivals. The SMEs, who are usually subcontractors, can rely on their set of specialised skills in certain niche areas, such as tiling, water-proofing, glass making, and installation services, in order to distinguish themselves from their rivals in the industry.

Hence, to reiterate, RBV is an efficiency-based explanation for performance differences (Barney, 1991; Conner, 1991; Peteraf & Barney, 2003; Teece, Pisano, & Shuen, 1997), where a firm's resources must be organised to capture value. In addition, they should be valuable, rare, and imperfectly imitable or substitutable in order to be source of a SCA (Barney, 1991). This would eventually guide the firm to make critical choices, which will form the basis of their strategy to achieve SCA (Porter, 1991) in a highly-competitive 'red-ocean market'.

4.1.3 Critique of the RBV Theory

Nonetheless, having introduced the RBV theory to account for the occurrence of prevalent subcontracting practices in Singapore's construction industry, various researchers have cautioned about the shortcomings in the RBV approach. Kraaijenbrink, Spender, and Groen (2009) have tried to review and assess the principal critiques evident in the RBV literature, arguing they fall into eight categories, of which, one of the categories caught the attention of the research team and warrants scrutiny due to its limited applicability. These eight categories are summarised in Annex 1. Other relevant critiques have also been discussed below.

Critique: RBV's applicability is too limited

It was mentioned that one limitation is the applicability of the RBV. It applies only to firms striving to attain SCA. For firms satisfied with their competitive position, the RBV does not bring much insight, as its relevance directly follows from managers' aspirations and intentions.

In addition, Barney (2002) indicated an important limit to the applicability of the RBV. It holds only as long as the 'rules of the game' in an industry remain relatively fixed. In unpredictable environments in which new technologies and/or new markets emerge and the value of resources can drastically change, there is a need to go beyond the RBV to explain a firm's SCA.

Critique: RBV Theory is too firm-centric and neglects surrounding environments

Porter (1991) commented that the RBV theory is most introspective and centred on the firm itself, thereby implying that the external factors might affect their goal of achieving SCA. Moreover, the assumed heterogeneity and immobility are not, however, sufficient conditions for SCA, according to what is expected from the RBV approach (Bridoux, 2004). This is discussed by Peteraf (1993); she presented four conditions underlying SCA: superior resources (heterogeneity within an industry), ex post limit to competition, imperfect resource mobility, and ex ante limits to competition. This relates to the fact that the RBV theory itself does not provide sufficient coverage to explain how factors from external environments influence a firm's ability to achieve SCA, such as an economic downturn, where for instance, the construction industry in the UK suffered a 39% decline in output in 1993 due to an economic downturn (Latham, 1994). Therefore, not only must the firm consider its own resources but it must also take into account its surrounding environment when formulating an appropriate and realistic strategy to achieve SCA in the long run, based on the choices made by the firm. This will also help to explain the issue of rampant subcontracting in the local construction industry in a more holistic manner.

4.1.4 Complementary Theory with the RBV Approach

The scrutiny and assessment of RBV have also pointed to a number of unresolved problems in the RBV approach. As such, some of these problems justify the adopted approach, such as to integrate the RBV theory and the firm's surrounding competitive environment. Bridoux (2004) suggested that the exercise of market power can help to build and protect the competitive advantage. Porter (1991) adopted a similar definition of competitive advantage, which facilitates the integration of both perspectives (Bridoux, 2004).

This is also supported by Barney (1991) and Rothaermel (2012), who, based on the chart in Fig. 4.3, recommended that another theory must be combined with the RBV theory to maximise a firm's success by achieving and sustain competitive advantage. It is called industry effects.

As depicted in Fig. 4.3, from 30 to 45% of superior organisational performance can be explained by firm effects (RBV) and 20% by industry effects (I/O view aka competitive environment). In addition, numerous RBV authors (Amit & Schoemaker, 1993; Barney, 1991; Conner, 1991; Mahoney & Pandian, 1992; Peteraf, 1993; Peteraf & Barney, 2003; Wernerfelt, 1984) have recognised that the resource-based perspective and industrial organisation tools, such as Porter's Five Forces Model, complement each other in explaining the sources of firm

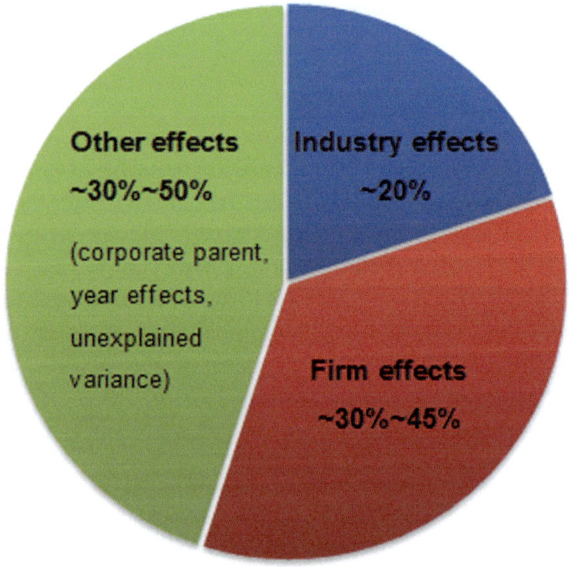

Fig. 4.3 Chart on factors affecting organisational performance

performance (Bridoux, 2004). Foss (1996, p. 19) summarised the thematic complementarities between Porter's framework and the RBV. The RBV is helpful in ascertaining the dangers of future competitive imitation through an analysis of the resources and capabilities of competitors. Porter may add an understanding of the external environment with concepts such as commitment, the role played by exit barriers, etc. (Bridoux, 2004). Besides thematic complementarities, there are also conceptual complementarities. Spanos and Lioukas (2001) presented two such similarities: (1) the RBV perspective and Porter's (1980) framework share the view that persistent above-normal returns are possible, and (2) both perspectives seek to explain the same phenomenon of interest (i.e., SCA). In addition, both perspectives assume that managers are rational and that a firm's ultimate goal is to increase its performance.

4.1.4.1 Proposed Complementary Theory— Porter's (1980) Five Forces Model

The adoption of Porter's Five Forces Model is an appropriate choice to complement the RBV theory, given that there are thematic and conceptual complementarities. In addition, the Five Forces Model has been a most influential model and has propelled strategic management to the very core of the management agenda. Therefore, the RBV theory of the firm is extended by incorporating Porter's Five Forces Model to give an overview of the connections between resources and competition that closely intertwine each other, which better reflects the actual realities faced by firms in the market. Hence, the extended RBV theory suggests that firm's resources provide the basis for strategy. Strategy should allow the firm to best exploit its resources relative to the competitive environment in order to achieve SCA over their rivals (Bridoux, 2004).

4.2 Porter's (1980) Five Forces Model

Professor Michael E. Porter (1980) first conceived of the Five Forces Model in the late 1970s. It is a simple but influential tool for the identification of where power lies in certain business circumstances by employing the outside-in perspective (Johnson, Scholes, & Whittington, 2008). The framework, as shown in Fig. 4.4, identifies five forces in the microenvironment that drive competition and threaten a company's ability to make a profit. Porter (1980) then defined the forces that drive competition, contending that the competitive environment is created by the interaction of five different forces acting on a business.

Moreover, Porter (1980) suggested that the competition intensity is governed by the relative strengths of these forces. By understanding the nature of each individual force, organisations will gain crucial insight to enable them to formulate the appropriate strategies to be successful in their markets (Ščeulovs & Gaile-Sarkane, 2010). This is how the attractiveness of the market and its overall profitability can be defined by the market structure (Slater & Olson, 2002). Then, the market structure in turn influences the strategic behaviour of organisations (e.g., market success depends on the competitive strategy). Thus, organisational success is indirectly dependent on the market structure, where the 'awareness of these forces can help a company stake out a position in its industry that is less vulnerable to attack' (Porter, 1979a, b, c).

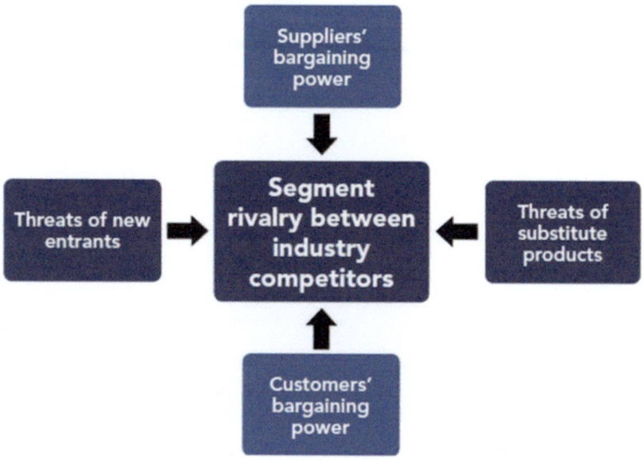

Fig. 4.4 Porter's five forces framework

4.2.1 Assumptions and Elaboration of the Five Forces Model

The five key forces that the model uses to identify and evaluate potential opportunities and risk (Karagiannopoulos, Georgopoulos, & Nikolopoulos, 2005) are explained in Table 4.1.

Table 4.1 Porter's (1980) five forces model

Assumptions		• Presence of many representative competitors • Three-stage chains made up of suppliers, rivals, and buyers • Accounts for potential entrants, substitutes, and direct rivals. • Attractiveness of an industry, in which a company operates, is determined by the market structure because market structure affects the behaviour of market participants (Raible, 2013)
Forces involved		Elaboration
1	Competitive rivalry	Rivalry intensity helps to determine the extent to which the value created by an industry will be dissipated through head-to-head competition. It is one of several forces that determine industry attractiveness
2	Threat of new entrants	Both potential and existing competitors' influence the average industry profitability. The key concept in analysing the threat of new entrants is the entry barriers. They can take diverse forms and are used to prevent an influx of firms into an industry whenever profits, adjusted for the cost of capital, rise above zero. As such, entry barriers exist whenever it is difficult or not economically feasible for an outsider to replicate the incumbents' position. The most common forms of entry barriers, except intrinsic physical or legal obstacles, are usually the scale and investment required to enter an industry as a credible competitor
3	Threat of substitutes	The threat that substitute products pose to an industry's profitability depends on the relative price-to-performance ratios of the different types of products or services to which customers can turn in order to satisfy the same basic need. The threat of substitution is also affected by switching costs, such as retraining, retooling, and redesigning, that are incurred when a customer switches to a different type of product or service
4	Bargaining power of suppliers	Supplier power is the mirror image of buyer power. First, supplier power typically focuses on the relative size and concentration of suppliers relative to industry participants. This is followed by the degree of differentiation in the inputs supplied. Hence, the ability to charge customers different prices is in line with differences in the value created for each of those buyers, which usually indicates that the market is characterised by high supplier power and at the same time by low buyer power (Porter 1989, 1996).
5	Bargaining Power of Customers	Size and the concentration of customers are the most important determinants of customer bargaining power (Karagiannopoulos et al., 2005). Other factors are the extent to which the buyers are informed and the concentration or differentiation of the competitors. As such, the bargaining power of customers helps influence the appropriation of the value created by an industry

4.2.2 *Applicability of Five Forces Model to the Local Construction Industry*

In addition, it is necessary to mention that the five forces have diverse degrees of effect in certain industries and that 'individual forces and their collective impact will change as the government policies and macroeconomic and environment conditions change' (Mohapatra, 2012).

4.2.2.1 Implementation of Labour Restricting Policies in Singapore

For instance, the Ministry of Manpower made changes to the labour restructuring policies to improve the image of the construction industry and to achieve productivity-led growth. This is to curtail reliance on foreign labour and to rebrand the construction industry as an efficient, modern, and non-labour-intensive industry. However, this has caused an arrest of growth in the number of foreign workers in Singapore, and many contractors face manpower concerns (Heng, 2015). As such, the construction industry has been struggling to meet the increasing construction demand. Key labour restructuring policies are introduced as follows:

- Foreign worker levy (FWL),
- Dependency ratio ceiling (DRC), and
- Man-year entitlement (MYE) quota.

4.2.2.2 Effects of Labour Restricting Policies in Singapore

The implementation of labour policies has had an immense effect on all construction firms for both large firms and SMEs within the industry. The various implemented labour policies and their associated effect on the firms will be discussed in Table 4.2.

Nonetheless, construction SMEs will bear the brunt of the effect due to the introduction of labour restructuring policies. As quoted by Chai Wai Fook, tax partner at Ernst & Young:

> Companies do recognise that there is no choice. The government is not changing direction. Either you innovate and improve productivity or you may be phased out. He said the changes will have a limited effect on the bigger companies; with the muscle to pay higher salaries to local workers, these companies become eligible to hire more foreign ones. Smaller firms may not be able to do this. In the long term, some smaller companies might drop out, simply because they can't compete, or they may be acquired, or they may acquire competitors so they can amalgamate resources and compete for government projects. (Cited in Tan, 2015 at http://www.businesstimes.com.sg/government-economy/singapore-budget-2015/budget-helps-construction-sector-but-challenges-remain)

Table 4.2 Labour restructuring policies

Objectives		The implementation of such policies is part of the restructuring effort that supports the national goals of (Ministry of Manpower, 2011): • Encouraging productivity improvement, • Reducing reliance on low-skilled foreign worker, and • Fostering inclusive growth	
Key policies involved		Elaboration	Associated effect
1	Dependency ratio ceiling (DRC)	A protectionist measure of the local population to ensure work for locals is still available in the construction industry, where it limits the maximum allowed ratio of foreign workers to the total workforce that a company in a specific industry is allowed to hire (Weng, 2013)	The DRC limits the number of foreign workers a firm can employ by raising the threshold salary for part-time workers so that employers do not hire local workers on token salaries to recruit more foreign workers (Weng, 2013)
2	Man-year entitlement (MYE)	A form of a work permit allocation system meant for construction workers from non-traditional source countries and the People's Republic of China (PRC) (MOM, 2012)	A decline of 45% of the MYE further shrinks the available pool of foreign workers for employers, which was a drastic drop within four years (Heng, 2015)
3	Foreign worker levy (FWL)	A form of financial levy that a company pays the government for every S Pass or work permit holder that it hires. The main aim of such a pricing mechanism is to control the foreign worker numbers in Singapore (CPFB, 2014)	An inevitable increase in the firm's capital as the FWL has been increased to limit the number of foreign workers in Singapore. It is intended to discourage the reliance on labour and encourage contractors in the construction industry to adopt more capital intensive and productive methods of construction
Overall effect		All construction firms will experience a manpower crunch and higher operating costs as a direct result of these policies	

Even with the recent lowering of MYE waiver levies, it would make it easier for companies to hire higher skilled workers, but the increase in the basic tier levy for basic skilled workers—from S$550 to S$650 in 2016, and subsequently to S$700 the following year—would see fewer basic skilled (R2) workers hired, doing nothing to alleviate the current labour crunch (Channel News Asia, 2015). This has evoked mixed reactions despite having a one-year respite from the increase in FWL after its announcement for the budget of 2015.

4.2.2.3 Analysis Using the Five Forces Model

Using the context of the implemented labour policies, a form of external factor that influences the nature of competition within the local construction industry via the five forces in the microenvironment, influences the way in which the local firms compete. This drives competition and threatens a company's ability to make profit (Dälken, 2014; Ščeulovs & Gaile-Sarkane, 2010). As such, Table 4.3 will present an analysis of the Five Forces Model regarding how such external factors will have an effect on the local industry that spurred firms to look towards subcontracting (Betts & Ofori, 1992).

From Table 4.3, it is clear that each of these five forces is directly relevant to construction. Tatum (1988) went as far as to suggest that construction has a high fit with each component of the Five Forces Model. Reiterating the earlier point that the market structure, such as the adoption of the MLSS in the local context, in turn influences the strategic behaviour of construction firms. Thus, organisational success is therefore indirectly dependent on the market structure by the implementation of an appropriate competitive strategy, such as the prevalent use of subcontracting practices from one firm to another or to a few other firms in completing work.

4.2.3 Critique of Porter's Five Forces Model

See Table 4.4.

4.2.4 Complementary Framework Review (RBV Approach and Porter's Five Forces)

As mentioned earlier, the RBV theory of the firm is extended by incorporating Porter's Five Forces Model to give an overview of the connections between resources and competition that closely intertwine each other. An example of a complementary framework proposed by Bridoux (2004) integrates both resources and the competitive environment as sources of performance, drivers of strategy, and, particularly, competitive behaviours. It was first developed to focus on the issues of performance and competition in a given (product and/or geographic) market. This framework, as shown in Fig. 4.5, is summarised as follows.

As depicted in Fig. 4.5, the resources and competitive environment affect the firm strategy. The firm strategy and performance in turn affect the competitive environment and resources, and all these changes generate new information, which in turn creates new learning opportunities and may lead to the creation and development of new resources. This constitutes a form of strategy creation, where the ongoing sequence of actions and reactions conditioned by the firm resources

Table 4.3 Analysis using Porter's Five Forces Model

Context		With the implementation of key labour restructuring policies, all construction firms have experienced a manpower crunch and higher operating costs	
Forces Involved		Explanation	Effect on local Industry
1	Threat of new entrants	The threat of new entrants is not potent within the construction industry, as the ease with which new firms can enter with minimal investment in Singapore has been noted. However, increasingly large investments by major Japanese construction firms in computing and engineering technology are a barrier to new entrants in the larger projects and high-technology end of the construction sector (Betts & Ofori, 1992)	An emerging entrance barrier is the direct effect of the labour restructuring policies, where firms will have to deal with higher operating costs and the increasing difficulty in acquiring manpower. This will erode potential profits and lengthen the break-even time for newcomers, which poses a higher psychological entry barrier, where the thought of being unable to achieve decent returns will deter newcomers from the already competitive industry
2	Threat of substitutes	As there are many firms of varying sizes and capacities already operating in the industry, the threat of substitutes remains. Moreover, the potential for substitute products from these firms gives scope for competitive advantage within an industry (e.g., the emergence and acceptance of precast components, such as prefabricated precast volumetric construction, to achieve high build ability and constructability scores by many firms)	Such occurrences resonate with an industry-wide effort by the government to promote sustainability in construction
3	Threat of suppliers' bargaining power	In the construction industry of each country, there are often many different suppliers of many products and services. The highest competitive advantage is possible when few enterprises dominate supply, where there is no competing product and when the supplier holds a threat of forward integration over the buyer (Betts & Ofori, 1992)	In Singapore's context, suppliers of raw materials and labour are usually abundant, and construction firms (buyers) generally have a greater say than the suppliers. Hence, firms that supply such resources will always compete vigorously and usually at the losing end, and hence are unable to achieve high levels of competitive advantage
4	Threat of customers' bargaining power	There are many opportunities for large construction enterprises to exploit this opportunity with material suppliers and specialist subcontractors. They are the customers/clients with the greatest competitive advantage when they are a large volume buyer, as they purchase undifferentiated products that are price sensitive but not quality sensitive to processes	The emergence of design-and-build and turnkey contractors are examples of the extent to which competitive advantage may be exploited in this way. However, such an arrangement is rare in Singapore. Nonetheless, due to large quantity purchases, large construction firms have significant bargaining power over suppliers

(continued)

Table 4.3 (continued)

Context		With the implementation of key labour restructuring policies, all construction firms have experienced a manpower crunch and higher operating costs	
Forces Involved		Explanation	Effect on local Industry
5	Threat by intense rivalry among competitors	Parts of the construction industry, especially at the lower end of the technological scale, do have large numbers of similarly sized competitors with undifferentiated products and services. This gives the greatest scope for competitive advantage when there are many competitors of equal size, slow industry growth, undifferentiated products or services, and low fixed costs. Coupled with low exit barriers, it causes a high entry and attrition rate among construction enterprises (Porter 1979a, b, c)	Jockeying for position among local construction firms is a strong competitive force within the construction industry. Not only has there been an increase in the number of employees within the industry but also the number of firms, especially small firms, has increased over the past decade, which shows the immense 'red-sea' rivalry in the local market

and competitive environment become exogenous events in the environment of other firms. The term 'firm' in the following section refers either to a single product firm in a single geographic market or to a business unit with the same characteristics.

With that, it lays out the plan as to how firms strive to achieve SCA via the complementary framework proposed by Bridoux (2004), as it addresses the shortcomings of the RBV theory by incorporating external forces that influence competition among firms in a competitive market using Porter's Five Forces Model. This is despite the fact that Porter's five forces is a rigid model that does not promote flexibility in terms of firm collaboration and is a static model that does not take into account changes in industry structures. Such flaws have made the Five Forces Model outdated.

4.2.4.1 Viability of Porter's Five Forces Theory and RBV Approach

Nonetheless, there is little evidence of construction enterprises systematically examining these five forces and their relevance to the strategic planning and management of their firms in the long term. Moreover, Porter's Five Forces Model falls short in explaining how firms, especially SMEs, would adapt to such challenges in order to survive in such trying times. This is despite complementing it with the RBV approach to explain how firms behave as they factor in their resources, relative to the competitive environment they are facing, which affects their strategy based on choices. It poses the question of whether, under such tough circumstances, the survivability of SMEs is at stake or at risk of elimination by larger firms. For instance, subcontractors who were in the lower rungs of the MLSS hierarchy have struggled to survive as their profit margins are successively cut,

Table 4.4 Criticisms of Porter's Five Forces Model

Factors		Elaboration	Outcome
1	No justification for the choice of the five environmental (external) forces	The validity for Porter's choice of the five forces cannot be proven (O'shaughnessy, 1984; Speed, 1989)	A trivial point to make, as it is the basis for the Five Forces Model. Insufficient evidence and magnitude to derail credibility of the model exist
2	The model only generates snap-shots	According to Thurlby (1998), Porter's Five Forces Model is static and does not consider time. Thus, it is more difficult to determine markets with higher competition dynamics due to rapid changes as time passes	Disadvantaged managers are unable to factor in the 'time dimension', which would enable managers to better consider market trends and changing environments. It is an inadequate and unrealistic model
3	The five forces framework does not guarantee a competitive advantage that is inviolable and sustained (Aktouf, 2004)	The five forces framework is a static model that does not constantly include changes in competitive environments (Karagiannopoulos et al., 2005). According to Hill and Jones (2008), firms must protect against the five forces, and it is increasingly important to collaborate and continuously innovate in the rising prominence of information technology (Karagiannopoulos et al., 2005; Holm, Eriksson, & Johanson, 1996)	The Five Forces Model fails to account for collaboration, as the model strictly prohibits such moves by firms, as they were initially assumed to compete against each other. It can be deemed an increasingly irrelevant model that does not fit with the changing times
4	Missing attention to key global factors, such as digitalisation, globalisation, and deregulation that have changed industry structures during the last decades	The criticisms made by Flower (2004) and Downes (1997) alluded that these three factors are key reasons for the changing industry structures. Grundy (2006) also mentioned that the model is not applicable to all markets	This is clear evidence that the model is unable to adapt to changing times to incorporate new global developments. We can conclude that the model does not provide a holistic perspective and that urgent modifications need to be made to the model to minimise its extinction
5	The model does not assess the resources and capabilities of a company	The model is too focused on the inner-workings/dynamics of the five forces that influence the firms' decision on their strategy. Little or no emphasis is made to factor in the resources and capabilities of the firms as they are relevant for analysing the overall profitability of the firm (Rivard, Raymond, & Verreault, 2006)	Although a limitation, its severity does not warrant attention, as it can be easily overcome with the RBV approach that factors in resources and capabilities of the firm. This is evidence that the model needs to be complemented with other suitable theories like the RBV to ensure maximum effectiveness

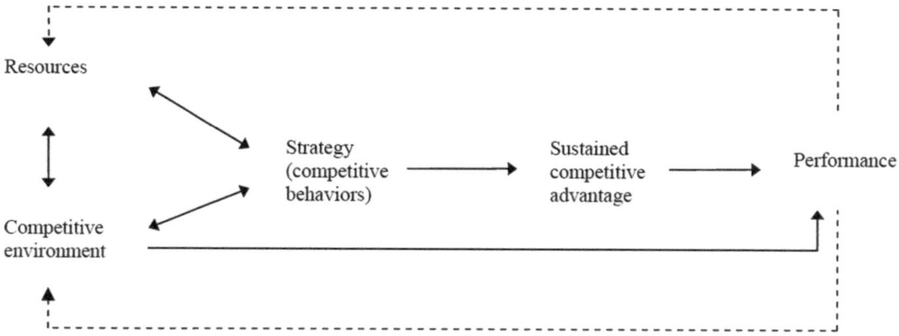

Fig. 4.5 Integrated framework of both resources and competitive environment

down the tiers of the MLSS hierarchy to a point where the lowest tier subcontractors have barely enough to cover the increased cost of hiring foreign workers. As such, there is a need to further examine other literature to enable an appropriate explanation for the rampant subcontracting practices that are occurring in Singapore's construction industry.

4.3 Flexible Firm Model

A highly probable model that could be adopted to explain the significance of subcontracting practices that are prevalent in the local industry would be the flexible firm model.

4.3.1 Current Industry Climate in Singapore

The adoption of the MLSS has been common in Singapore, coupled with the fact that there has been an approximately 47% increase in the total number of firms in the industry from 1996 to 2013, and more notably a 28% increase in the number of small firms from 2005 to 2013 is shown in Tables 4.4 and 4.5, respectively. This implies that not only is the construction industry booming, the presence of many small firms indicates that there are multiple levels of ongoing subcontracting. This current industry development runs contrary to what has occurred during most of the twentieth century, where SMEs were considered to be a declining sector and archetypical in which "informal" and "pre-modern" labour relations and technologies have hindered the process of economic development (Taymaz & Kilicaslan, 2002). As such, "subcontracting" was considered a form of domination of large firms over small ones where large firms benefited from low wages and flexible work arrangements in small firms, in the context of the construction

Table 4.5 Kinds of flexibilities and types of workforce

Purpose		A strategic model of labour utilisation used to recognise the need for organisations to have enhanced flexibility to meet the ever-evolving market and competitive pressures	
Flexibility types		Elaboration	Associated effect
1	Functional flexibility	This enables the employer to quickly and smoothly transfer employees between activities and tasks (Atkinson, 1984) (e.g., moving multi-skilled workers between different skilled jobs or moving workers between indirect and direct production jobs)	This will result in the ability of the same labour force to change when products and production methods change in the short and medium term (Atkinson, 1984)
2	Financial flexibility	Enables pay flexibility, where a shift to a new pay and remuneration system that facilitates either numerical or functional flexibility (Atkinson & Meager, 1986) (e.g., the use of assessment-based pay systems in place of job rate systems). It also allows a continued shift to the plant level bargaining and widening differentials between skilled and unskilled workers as employees' salaries and other employment costs change with the state of supply and demand in the external labour market (Atkinson & Meager, 1986)	The significance lies more in the ability to accrue competitive advantage required in the long run and when market conditions change
3	Numerical flexibility	Enables the firms to have the ability to adjust the level of labour to meet fluctuations in demand. Essentially, at any time, the numbers employed would exactly match the numbers required. Therefore, it facilitates the smooth implementation of various hiring and firing policies (Atkinson, 1984)	For short-term changes in the demand for labour, numerical flexibility allows headcount to be quickly and easily changed. However, it results in a looser contractual relationship between managers and workers. It can be hard for owners to exert control and supervision when necessary
Types of workforce		Elaboration	Associated effect
1	The core workforce	Characterised by permanent, highly skilled employees with internal career paths (Wood, 1989; Procter et al., 1994), as they experience a higher degree of job security with resources provided for training in firm-specific skills not readily acquired. This is a form of functional flexibility (Burgess, 1997; Hakim, 1987)	This results in the ability to provide organisations with functional flexibility. They are a pool of skilled full-time employees that could be mobilised in response to the emerging technological, market, or product changes. This is beneficial and desired for firms
2	The peripheral workforce	The key function or strategic aspect of this workforce is to undertake day-to-day activities that are important but not vital to the organisation (Atkinson, 1984)	This affects the peripheral groups: part-time, temporary, or contractual workers, as firms deal with market fluctuations, causing labour demand changes. There are growing calls to address the fact that part-time/temporary workers might, numerically and strategically, constitute the core component of a company's workforce (Walsh, 1993)

industry. Therefore, it was thought that the SME sector would be eliminated by more efficient and advanced large firms (Taymaz & Kilicaslan, 2002) but was proved otherwise. As such, the flexible firm model is introduced to explain the prevalence of subcontracting practices, its effects, and the various challenges that have motivated firms to adopt such practices in Singapore.

New pressures that are affecting the employment plans of the majority of firms are (1) market stagnation, (2) economic uncertainty, (3) technological change, and (4) reduction in working hours. Thus, employment practices are shifting in response to this need for flexibility (Atkinson, 1984) as companies are seeking flexibility in the functional, numerical, and financial aspects. For instance, as reiterated earlier, the government's imposition of labour restricting policies as part of its strategy to achieve productivity-led growth has caused numerous implications on manpower requirements on many firms in a labour-intensive industry. As such, many firms are forced to adapt rapidly to such changes or face elimination. Hence, the key to their survival is flexibility.

4.3.2 Justification for the Adoption of the Flexible Firm Model

To enable a more robust analysis of changes in a workplace organisation, flexibility is often factored in, especially to analyse labour utilisation (Broadbent, 2002). Atkinson's flexible firm model (1984) is an example of how theorists have characterised these changes, as it focuses on ways in which the greater use of an adjustable periphery, less rigid work practices, and implementation and utilisation of new technology may contribute to a firm's profitability and competitiveness (Hunter, McGregor, Macinnes, & Sproull, 1993). It was observed that employers have created a large periphery of non-permanent employees to bear the burden of restructuring.

The two main approaches to flexibility, as proposed by Guerrier and Lockwood (1989), are the following:

1. To develop a core of permanent full-time employees trained in a variety of skills to move from function to function as demands require.
2. To rely on part-time, casual, and short-contract staff to provide the necessary variation in the number of employees required. Usually these employees would only have limited skillsets.

Moreover, the flexible firm model can be viewed as a new strategic, innovative tool in response to dealing with a firm's changing environment. The typical changes in a firm's environment include economic change, unemployment, legal restrictions on union actions and most importantly global competition (Ofori & Debrah, 1998). The key characteristic of the flexible firm model is that it emphasises the core-periphery workforce, where it allows one to design the workforce to proactively meet its business needs through flexible staffing arrangements. To enable easier comprehension, a visual illustration of the flexible firm model is shown in Fig. 4.6. Its components are explained in the subsequent sections.

In addition, the extensive use of subcontracting in Singapore is evidenced by the interview results (see Chaps. 7 and 8) conducted by the research team, which indicates that the main contractors are adopting employment approaches similar to those portrayed in Atkinson's (1984) flexible firm model. This is an area for concern, as such subcontracting practices are a form of human resource (HR) management. If the firm is unclear on the long-term business needs for labour flexibility, staffing activities are often short-term reactions to product or market pressures. This can result in poor employee relations, reduced quality, and reduced productivity. As such, this model warrants further in-depth study to assess whether this model is the key behind the rampant subcontracting that has been ongoing and how it has had an effect on Singapore's construction productivity.

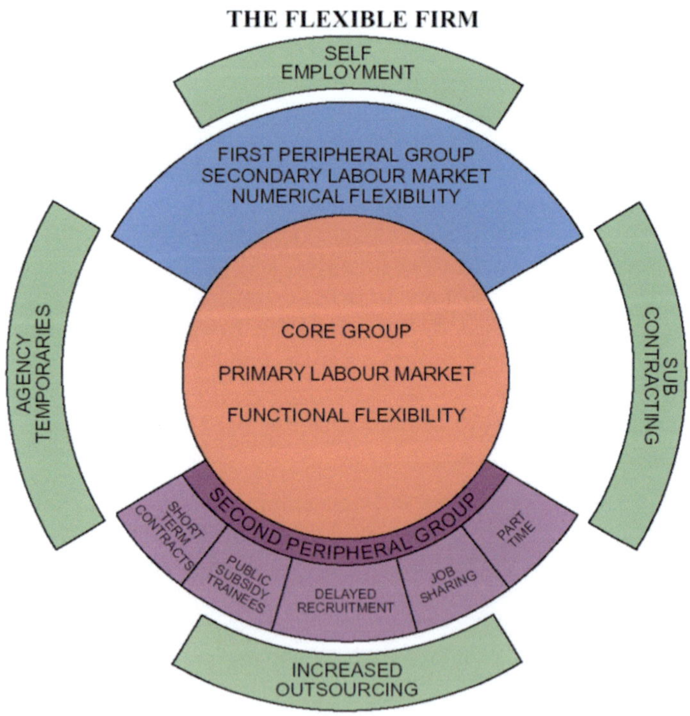

Fig. 4.6 Flexible firm model

4.3.3 Features of the Flexible Firm Model

In acknowledging that organisations will require enhanced flexibility to meet the ever-evolving market and competitive pressures, Atkinson (1984) proposed the concept of a flexible firm. This is a strategic model of labour utilisation that is becoming increasingly common as firms face restructuring to meet a number of challenges, such as market uncertainty, cost pressures, and technological change (Atkinson & Meager, 1986). Atkinson pointed out that there are three different kinds of flexibility, namely, functional flexibility, financial flexibility, and numerical flexibility.

4.3.4 Types of Workforce

The creation of the flexible firm model provided the means to analyse the move towards greater use of contingent labour, the adoption of employment strategies such as in sourcing and outsourcing, and the concentrated use of compulsory competitive tendering in the public sector. More importantly, it provides a framework for focusing on the extent of change and the development of new work patterns at the firm level, providing a more complex and useful means of analysis (Procter et al., 1994).

In addition, utilising the concepts of core and periphery, the segmentation of the workforce could be analysed (Atkinson, 1984), where the model could offer a starting point for examining the segmentation of Singapore's construction labour market and provide direction in the exploration of why subcontractors are predominant in the construction labour workforce. The model employs the distancing strategy, which requires the replacement of standard, traditional employment relationships with commercial ones. Table 4.5 summarises the various kinds of flexibilities and types of workforce (Table 4.6).

4.3.5 Applicability of Flexible Firm Model to Local Construction Context

The adoption of the flexible firm model allows for greater use of subcontractors in the Singapore construction industry (Ofori & Debrah, 1998) where the majority of the employees are self-employed, part-timers, and workers for subcontractors. Most important of all, the model is applied to reduce costs in changing global markets, resulting in financial flexibility. This is desired for the construction SMEs where the model offers employees both functional and numerical flexibility (Ofori & Debrah, 1998). This is in addition to the lower operating, overhead, and fixed costs that these SMEs would enjoy with the application of the model. As such, to ensure the

Table 4.6 Key criticisms of the flexible firm model

Issues		Elaboration	Outcome
1	No empirical evidence at an aggregate level to support significant changes in employers' use of HR strategies	Pollert (1988) considered 'flexibility' to be a neat and bold model fraught with methodological and conceptual problems	It is difficult to determine firms' HR strategies due to the difficulties in acquiring such sensitive information. Nonetheless, it is still a decent starting point to analyse firm's HR strategies
2	The tendency to see flexibility as good and rigidity as bad is far too simple	Flexibility simultaneously involves advantages as well as tensions and conflict (Geary, 1992)	A more holistic approach is required to enable the comprehensiveness of the model to factor in more scenarios and circumstances to achieve better analytical results from the model. This is not severe enough to derail the credibility of the model
3	Some firms have adopted 'anti-core-periphery' strategies that run contrary to the flexible firm theory	Firms have abandoned the use of peripheral workers after some unpleasant experiences (Hakim, 1990)	This is similar to Geary's (1992) assertion that the widespread use of flexibility has not resulted in significant benefits to many firms. This casts doubt on the reliability of the flexible firm theory

effectiveness of the model, it requires a sophisticated interplay of strategic business planning and HR management (Ofori & Debrah, 1998). Hence, the flexible firm model has rekindled interest in research on subcontracting as part of employers' labour-use strategies (Harrison & Kelley, 1993; O'Reilly, 1992a, b).

Moreover, the flexible firm model provides an early and innovative framework to analyse the move towards greater use of contingent labour and the adoption of employment strategies, such as in sourcing and outsourcing. It also offers a platform to understand the segmentation of Singapore's construction labour market and provides direction in the exploration of why subcontractors are predominant in the construction labour workforce, as can be seen in the MLSS that is common in Singapore's construction industry. As such, with the aid of the flexible firm model, one can better appreciate the motivations of firms that employ subcontracting. To reiterate, results from the interviews with main contractors and subcontractors and the subsequent analysis in Chaps. 7 and 8 concur with the theories adopted in the flexible firm model, as they reveal the underlying motivations for construction firms to conduct subcontracting.

4.4 Formulation of a Combinative Explanatory Framework

The combination of the three theories, in the form of a combinative explanatory framework, serves to merge the literature into a common platform that will serve to exemplify the reasons for the presence of the MLSS in the local industry, which has led to the prevalence of subcontracting. Such a theoretical framework would also cause some significant effects in the industry, more notably, in the area of construction productivity, which has been a discussion point since the recent labour productivity decline of −3.0% in 2013 (Building and Construction Authority, 2016).

4.4.1 *Features of the Combinative Explanatory Framework*

The combinative explanatory framework is as shown in Fig. 4.7.

In Fig. 4.7, the framework depicts a single construction firm having to manage its own resources and competencies in order to achieve competitive advantage. However, most other firms are also employing the same tactic. In addition, as discussed in Chap. 3, there are many firms in the local industry, which ignites competition (represented in red, indicating a "red-ocean" market or highly-competitive market). Firms should strive to reach the "blue ocean market", where organizations can enjoy working in a marketplace that is free of competitors, thereby achieving sustained competitive advantage that is desired. Traditionally,

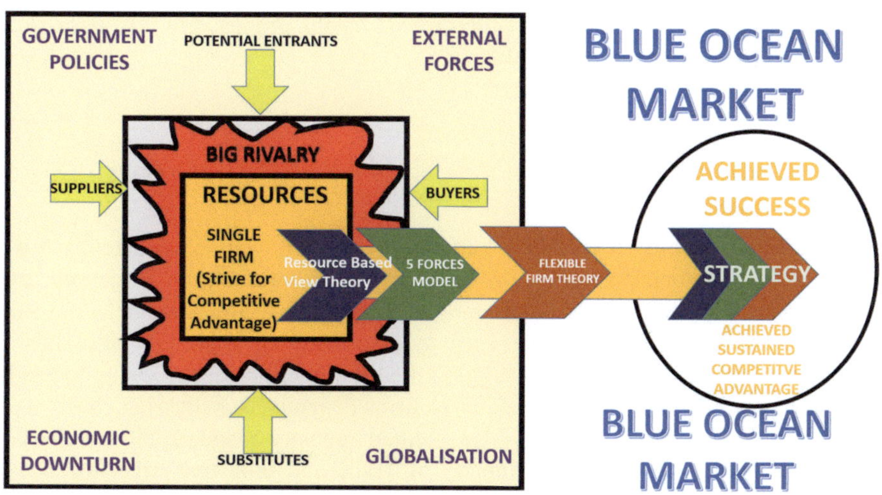

Fig. 4.7 Combinative explanatory framework

firms employ the RBV approach as an initial strategy to enable competition in the red-ocean market. However, as discussed earlier, RBV has a few glaring flaws in its theory and may not be adequate to explain the significance of subcontracting practices that are adopted by these firms. Hence, Porter's (1980) Five Forces Model has been brought into supplement the flaws in the RBV approach. The five forces are illustrated in the four yellow arrows and the red portions, which represent intense rivalry among firms in the red-ocean market. Nonetheless, despite complementing the RBV theory with Porter's five forces, valid criticisms of the five forces, such as its inability to account for changing industry structures due to the emergence of new global developments and the inability to factor in collaboration with the changing times, threaten to nullify the reasons subcontracting practices were adopted. As such, the flexible firm theory is brought into account for the ever-evolving market and competitive pressures that the firms face. Such pressures are caused by changing market conditions from external forces beyond the firm's control, such as changes in government policies, economic downturn, globalisation, and other unspecified external forces, which are labelled in purple. Therefore, by employing the flexible firm theory, the firms can overcome these macro-forces by employing enhanced flexibility in their labour utilisation to ensure survivability. This is represented by the brown arrow that depicts firms that "break-through" or overcome the boundaries caused by macro-forces.

Nonetheless, should firms want to achieve SCA in the long run, an appropriate strategy is required, as represented by the combination of the blue, green, and brown arrows. Firms will enjoy eventual success in a blue ocean market, where firms avoid intense rivalry from many firms. As subcontracting is not just a typical practice but is a strategy that is prevalently adopted in the industry, this combinative explanatory framework proposed by the research team seeks to explain the reasons from a theoretical perspective first, using various available literature to explain its prevalence. In subsequent chapters, the analysis will be conducted from the fieldwork interviews that will seek to reinforce the reasons proposed in this theoretical combinative explanatory framework.

4.4.2 Credibility of Combinative Explanatory Framework

In trying to produce a plausible explanation for the rampant subcontracting done locally and its associated effects on construction productivity in Singapore, the above framework has been proposed by the research team. Nonetheless, some literature points to subcontracting as a system resulting from the large firm's supremacy over smaller ones and suggests that the SME industry would be eradicated by more efficient and advanced large firms (Taymaz & Kilicaslan, 2002) and would put an end to the extremely fragmented industry plagued with numerous small firms lacking in capacity (Ministry of Trade and Industry, 2010) that have caused extensive subcontracting. This has reduced the continuity of working relationships, rendering the segmented approach of project delivery and management inefficient and thereby a culprit in causing productivity decline. Meanwhile, other literature has proven that, in the Singapore context, SMEs have flourished and extended their influence as they offer numerous economic benefits to their customers and business partners (Ofori & Lim, 2009), which could imply that the SMEs, operating within an MLSS, may not be the root cause of productivity decline despite their thriving presence.

Therefore, to determine the extent to which the subcontracting practices and the MLSS have affected construction productivity locally, detailed interviews have been conducted to qualitatively determine whether this has been the case. The preparatory work and several considerations for the interviews will be discussed in the next chapter.

References

Aktouf, O. (2004). The false expectations of Michael Porter's strategic management framework. *Revista Universidad & Empresa, 6*(6), 9–41.

Amit, R., & Schoemaker, P. J. H. (1993). Strategic assets and organizational rent. *Strategic Management Journal, 14,* 33–46. https://doi.org/10.1002/smj.4250140105.

Atkinson, J. (1984). Manpower strategies for flexible organisations. *Personnel Management, 16* (8), 28–31.

Atkinson, J., & Meager, N. (1986). *New Forms of Work Organization*. London: IMS.

Barney, J. B. (1991). Firm resources and sustained competitive advantage. *Journal of Management, 17,* 99–120.

Barney, J. B. (1994) Bringing managers back in: A resource-based analysis of the role of managers in creating and sustaining competitive advantages for firms. In *Does management matter? On competencies and competitive advantage. The 1994 Crafoord lectures* (pp. 1–36). Lund, Sweden: Lund University, Institute of Economic Research.

Barney, J. B. (2002). *Gaining and Sustaining Competitive Advantage*. Upper Saddle River, NJ: Prentice Hall.

Betts, M., & Ofori, G. (1992). Strategic planning for competitive advantage in construction. *Construction Management and Economics, 10*(6), 511–532. https://doi.org/10.1080/01446199200000049.

Bridoux, F. (2004) A resource-based approach to performance and competition: An overview of the connections between resources and competition. *IAG Working Papers,* 2004/110.

Broadbent, K. (2002). Flexibility at work? The feminization of part-time work in Japan. *The Journal of Industrial Relations, 44*(1), 3–18.

Building and Construction Authority. (2016). *Total Economy and Construction GDP* (p. 1). Singapore: BCA Infonet.

Burgess, J. (1997). The flexible firm and the growth of non-standard employment. *Labour & Industry, 7*(3), 85–102.

Channel News Asia. (2015). Singapore News. Retrieved from Firms welcome break from foreign levy hike: http://www.channelnewsasia.com/news/singapore/firmswelcome-break-from/1676026.html.

Conner, K. (1991). A Historical comparison of resource-based theory and five schools of thought within industrial organization economics: Do we have a new theory of the firm? *Journal of Management, 17*(1), 121–154.

Cool, K. C. A., & Luis, I. D. (2002). Constructing Competitive Advantage, *Handbook of Strategy and Management*, Sage publications.

CPFB. (2014). *Foreign Worker Levy (FWL)*. Retrieved from CPF Board: http://mycpf.cpf.gov.sg/Employers/Employers_Guide_to_CPF/New_Employer/ne_Levies_and_donations/NewEmployer_LeviesandDonations_FWL.htm.

Dälken, F. (2014). Are porter's five competitive forces still applicable? a critical examination concerning the relevance for today's business. In: *3rd IBA Bachelor Thesis Conference*. University of Twente.

Downes, L. (1997). Beyond Porter. Context Magazine.

Flower, E. (2004). Competition, technology, and planning: Preparing for tomorrow's library environment. *Information Technology & Libraries, 23*(2), 67–69.

Foss, Nicolai. (1996). Research in strategy, economics, and Michael Porter. *Journal of Management Studies, 33*(1), 1–24.

Geary, J. F. (1992). Employment flexibility and human resource management: The case of the three American electronic plants. *Work, Employment & Society, 6*(2), 251–270.

Grant, R. M. (1996). Toward a knowledge-based theory of the firm. *Strategic Management Journal, 17,* 109–122.

Grundy, T. (2006). Rethinking and reinventing Michael Porter's five forces model. *Strategic Change, 15*(5), 213–229.

Guerrier, Y., & Lockwood, A. (1989). Core and peripheral employees in hotel operations. *Personnel Review, 18*(1), 9–15.

Hakim, C. (1987). Trends in flexible workforce. *Employment Gazette, 95,* 549–560.

Hakim, C. (1990). Core and periphery in employers' workforce strategies: Evidence from the 1987 E.L.U.S. survey. *Work, Employment & Society, 4*(2), 157–188.

Hamel, G., & Prahalad, G. K. (1994). *Competing for the Future*. Boston: Harvard Business School Press.

Harrison, B., & Kelly, M. R. (1993). Outsourcing and the search for "flexibility". *Work, Employment and Society, 7,* 213–235.

Helfat, C. E., & Peteraf, M. A. (2003). The dynamic resource-based view: Capability lifecycles. *Strategic Management Journal, 24,* 997–1010.

Heng, N. (2015). *Labour restructuring policies and their implications on the multi-layered subcontracting system in the Singapore construction industry* (Unpublished BSc. Building study). National University of Singapore, Singapore.

Hill, C. W., & Jones, G. R. (2008). *Strategic Management: An Integrated Approach*. Boston: Cengage Learning.

Holm, D. B., Ericksson, K., & Johanson, J. (1996). Business networks and cooperation in international business relationships. *Journal of International Business Studies (Special Issue Supplement), 27*(5), 1033–1053.

Hunter, L., McGregor, A., Macinnes, J., & Sproull, A. (1993). The 'flexible firm': Strategy and segmentation. *British Journal of Industrial Relations, 31*(3), 383–407.

Johnson, G., Scholes, K., & Whittington, R. (2008). *Exploring Corporate Strategy: Text and Cases*. London: Pearson Education.

Karagiannopoulos, G. D., Georgopoulos, N., & Nikolopoulos, K. (2005). Fathoming Porter's five forces model in the internet era. *Info, 7*(6), 66–76.

Kraaijenbrink, J., Spender, J., & Groen, A. (2009). The resource-based view: A review and assessment of its critiques. *Journal of Management, 36*(1), 349–372. https://doi.org/10.1177/0149206309350775.

Latham, M. (1994). Constructing the team: Final report, *Joint review of procurement and contractual arrangements in the United Kingdom construction industry*. London: HMSO.

Mahoney, J. T., & Pandian, J. R. (1992). The resource-based view within the conversation of strategic management. *Strategic Management Journal, 13,* 363–380. https://doi.org/10.1002/smj.4250130505.

Ministry of Trade and Industry. (2010). *Construction Working Group Report* (p. 94). Singapore: Economic Review Committee, MTI.

Mohapatra, S. (2012). IT and Porter's Competitive Forces Model and Strategies. In: *Information Systems Theory* (pp. 274). New York: Springer.

MOM. (2011). *Foreign Worker Levy Changes: Strong Signal to Businesses to Invest in Productivity Changes to take effect from January 2012*. Retrieved from Press Release Details: http://www.mom.gov.sg/newsroom/Pages/PressReleasesDetail.aspx?listid=349.

MOM. (2012). *Further Moderating Demand for Foreign Manpower*. Retrieved from Ministry of Manpower- Press Release Detail: http://www.mom.gov.sg/newsroom/Pages/PressReleasesDetail.aspx?listid=409.

O'Reilly, J. (1992a). Banking on flexibility: A comparison of the use of flexible employment strategies in the retail banking sector in Britain and France. *International Journal of Human Resource Management, 3*(1), 35–58.

O'Reilly, J. (1992b). Subcontracting in banking: Some evidence from Britain and France. *New Technology, Work and Employment, 10,* 7–15.

Ofori, G., & Debrah, Y. A. (1998). Flexible management of worker: Review of employment and practices in the construction industry in Singapore. *Construction Management and Economics, 16*(4), 397–408.

Ofori, G., & Lim, K. H. (2009). Multi-layered subcontracting system in Singapore's construction industry. In K. Kobayashi, A. R. Kharuddin, G. Ofori, & S. Ogunlana (Eds.), *Joint Ventures in Construction* (pp. 216–226). Great Britain: Thomas Telford Limited.

O'Shaughnessy, J. (1984). *Competitive Marketing: A Strategic Approach*. Boston: Allen & Unwin.

Peteraf, M. (1993). The cornerstones of competitive advantage: A resource-based view. *Strategic Management Journal, 14*(3), 179–191.

Peteraf, M., & Barney, J. (2003). Unravelling the resource-based tangle. *Managerial and Decision Economics, 24,* 309–323.

Pollert, A. (1988). The 'flexible firm': Fixation or fact? *Work, Employment & Society, 2*(3), 281–316.

Porter, M. E. (1979a). How competitive forces shape strategy. In: *Harvard Business Review* (pp. 21–38).

Porter, M.E. (1979b) How competitive forces shape strategy. In: *Harvard Business Review*, March-April, (pp. 137–146).

Porter, M. E. (1979c). The structure within industries and companies' performance. *Review of Economics and Statistics, 61*(2), 214–227.

Porter, M. E. (1980). *Competitive Strategy: Techniques for Analyzing Industries and Competitors*. New York: Free Press. (Republished with a new introduction, 1998).

Porter, M. (1991). Towards a dynamic theory of strategy. *Strategic Management Journal, 12*(2), 95–117. https://doi.org/10.1002/smj.4250121008.

Procter, S. J., Rowlinson, M., McArdle, L., Hassard, J., & Forrester, P. (1994). Flexibility, politics and strategy: in defence of the model of the flexible firm. *Work, Employment and Society, 8*(2), 221–242.

Raible, M. (2013). *Industrial Organization theory and its contribution to decision-making in purchasing*. (Unpublished Bachelor of Behavioural, Management and Social Sciences Thesis). University of Twente, Netherlands.

Rivard, S., Raymond, L., & Verreault, D. (2006). Resource-based view and competitive strategy: An integrated model of the contribution of information technology to firm performance. *Journal of Strategic Information Systems, 15*(1), 29–50.

Rothaermel, F. T. (2012). *Strategic Management: Concepts and Cases* (p. 5). Pennsylvania: McGraw-Hill/Irwin.

Ščeulovs, D., & Gaile-Sarkane, E. (2010) Competitive Advantage for Small and Medium Enterprises. In: *Business, Management and Education: Scientific Conference: Conference Proceedings* (pp. 1–10, ISBN 9789955287575. ISSN 2029–7963), Lithuania, Vilnius, November 18, 2010, Vilnius: Vilnius Gediminas Technical University.

Shirazi, B., Langford, D., & Rowlinson, S. (1996). Organizational structures in the construction industry. *Journal of Construction Management and Economics, 14*(3), 199–212.

Spanos, Y. E., & Lioukas, S. (2001). An examination into the causal logic of rent generation: Contrasting Porter's competitive strategy framework and the resource-based perspective. *Strategic Management Journal, 22*(10), 907–934.

Slater, S. F., & Olson, E. M. (2002). A fresh look at industry and market analysis. *Business Horizons, 45*(1), 15–22.

Speed, R. (1989). Oh Mr Porter! A re-appraisal of competitive strategy. *Marketing Intelligence & Planning, 7*(5/6), 8–11.

Tan, M. (2015). *Budget 'helps construction sector, but challenges remain'*. [online] The Business Times. Available at: http://www.businesstimes.com.sg/government-economy/singapore-budget-2015/budget-helps-construction-sector-but-challenges-remain. Accessed September 14, 2016.

Tatum, C. B. (1988). Technology and competitive advantage in civil engineering. *ASCE Journal of Professional Issues in Engineering, 114,* 256–264.

Taymaz, E., & Kilicaslan, Y. (2002). *Subcontracting dynamics and economic development: A study on textile and engineering industries*. Economic Research Centre Working Papers in Economics 01/08 (pp. 1–25). Ankara, Turkey: Economic Research Centre, Middle East Technical University.

Teece, D. J., Pisano, G. P., & Shuen, A. (1997). Dynamic capabilities and strategic management. *Strategic Management Journal, 18,* 509–533.

Thurlby, B. (1998). Competitive forces are also subject to change. *Management Decision, 36*(1), 19–24.

Walsh, J. (1993). Internalization vs. decentralization: An analysis of recent developments in pay bargaining. *British Journal of Industrial Relations, 31,* 409–432.

Weng, T. H. (2013). Economic growth and inequality in Singapore: The case for a minimum wage. *International Labour Review, 152*(1), 107–123.

Wernerfelt, B. (1984). A resource-based view of the firm. *Strategic Management Journal, 5*(2), 171–180.

Wood, S. (1989). *The Transformation of Work: Skill*. Unwin-Hyman, London: Flexibility and the Labour Process.

Chapter 5
Research Design and Methodology

Abstract This chapter presents the research methods adopted and its design in order to attain the aim and objectives of the study. It highlights the formulation of the data collection methods and explains the methods of analysing the collected data to determine whether subcontracting practices have an influence on construction productivity in Singapore. The primary and secondary data collected from the in-depth interviews and questionnaires were analysed.

This chapter explains the research method adopted to accomplish the objectives that were outlined in Chap. 1. It highlights the formulation of the data collection methods for this study and provides explanations concerning the adopted methods of analysing the collected data to determine whether subcontracting practices have indeed caused an effect on construction productivity in Singapore. The research team also factored in current and past research work that has been carried out on subcontracting practices within the construction industry and the nature of the local construction industry. Such primary and secondary data collection from interviews and questionnaires were analysed as part of a research framework as shown below. There are three stages as discussed below.

The first stage of the research is the identification stage. It consists of an extensive literature review on the topic of the prevalent use of subcontracting practices within the construction industry globally and the nature of Singapore's construction industry. With that, it enhances the understanding of the state of the construction industry both worldwide and locally, with knowledge on the types of subcontracting practices adopted within the industry. This will provide ample information for the design of the interview questionnaire. The aim is to investigate the root cause of the research problem and propose recommendations for future research work in this realm.

The second stage of the research is the analysis stage. It comprises the formulation of interview questions and an in-depth use of data mining, a form of data analytics used to analyse the root causes of the prevalence of subcontracting practices in Singapore's construction industry, by understanding the underlying reasons and motivations from key players in the industry. Subsequently, a

L. Sui Pheng and S. Jia Zheng, *Construction Productivity in the Multilayer Subcontracting System*, Management in the Built Environment,
https://doi.org/10.1007/978-981-13-1831-3_5

sequential mixed method using a qualitative/quantitative sequence has been employed, where it is a form of methodological triangulation to enhance the internal validity of the research findings (Muhammad & McGoldrick, 2003). The data collection has been conducted with both well-established main contractors and prominent subcontractors within the local construction industry. This is to maintain the balance of the results and enable applicability to the local industry. Although this may seem like a qualitative study, where the data collection is mainly through interviews, the comprehensiveness of the interviews, coupled with the analytical process of data mining will provide better insight to the true state of the local industry. Therefore, this study will be able to derive an insightful result from the analysis that will support the research hypothesis in this study.

The third and final stage is the findings stage. It utilises the results from the data analysis and interviews to verify the authenticity of the state of the local industry and the prevalence of the subcontracting practices. Most importantly, it also serves to determine the extent of the effect of such subcontracting practices on Singapore's construction productivity. Therefore, this study will be able to arrive at an insightful conclusion that is derived from the analysis. Lastly, recommendations regarding the research problem and future studies are made to improve the situation, for the betterment of the local construction industry. Figure 5.1 shows an illustration of the research design.

As shown in Fig. 5.1, the research design comprises the literature review process, where online journal articles will be utilised and key information will be extracted in a bid to explain the significant occurrence of subcontracting practices in the local industry.

5.1 Data Collection Methods

5.1.1 Qualitative Research

Qualitative research enables an insightful understanding into a research problem (Malhotra, 1996). This study will adopt a qualitative approach by means of in-depth interviews to gain greater insight on the state of the local construction industry and the underlying reasons for the prevalence of subcontracting practices utilised by both main contractors and subcontractors. Twenty in-depth interviews were conducted, where interviewees consisted of senior management personnel and owners from main contractor and subcontractor firms, respectively. Out of the 20 interviewees, half of them are company representatives from main contractor firms, with the remaining from subcontractor firms. These selected interviewees were also key players in the industry. Inexperienced contractors were deliberately kept out of the sample to prevent inaccuracy in the collected data. Due to privacy concerns, some interviews were tape-recorded if allowed, while others were manually written; all

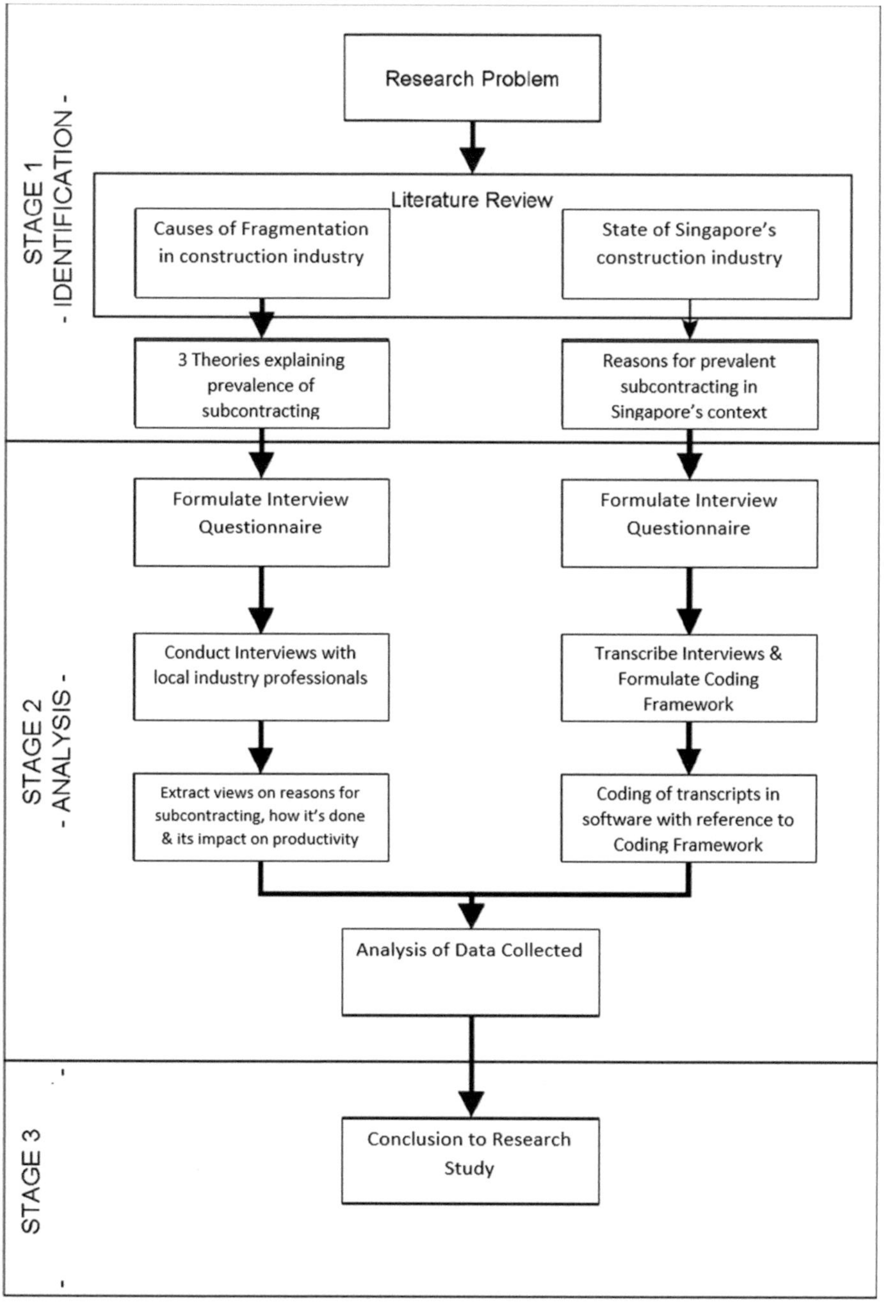

Fig. 5.1 Illustration of the research design

were transcribed for further analysis via data-mining software. The main issues that were discussed and recorded during the interviews are as follows:

Checklist of issues discussed and recorded during the interviews:

1. Rate of adoption of subcontracting practices,
2. Key reasons for subcontracting,
3. Selection requirements for subcontracting firms by contractors,
4. Opinions on the preference for core team of subcontracting firms,
5. Views on whether subcontracting practices aided their firm's productivity levels, and
6. Views on the current state of Singapore's construction industry.

All responses from the interviews were collated and input into a data-mining software to extract and identify undiscovered patterns and establish hidden relationships from the given responses. Subsequently, conclusions will be deduced from the results produced by the data-mining software, through the process of data analytics, to verify the study's hypothesis as mentioned in Chap. 1. Such a qualitative approach was adopted to enable a comprehensive understanding of the issues that interviewees face in the local industry with regards to subcontracting practices and the reasons for its widespread usage.

5.1.2 Justification for Interviews

Interviews were identified to be the most appropriate means of collecting qualitative data to understand contractors' preferences for subcontracting. Such interviews would allow respondents the freedom to express their thoughts and opinions. Moreover, such exploratory interviews enabled increased information gathering on the underlying motivations for the adoption of subcontracting and its prevalence in the local industry. Oppenheim (1992), in his book *Questionnaire Design, Interviewing and Attitude Measurement*, stated that the exploratory interview is "concerned with trying to understand how ordinary people think and feel about the topic of concern to the research". During the interviews, both factual and attitude questions were asked to enable a holistic understanding of the prevalence of subcontracting practices in Singapore. Attitude questions consist of the bulk of the interview questionnaire. Oppenheim (1992) also wrote that "we should not reply on single questions when we come to measure those attitudes that are most important to our study; we should have a set of questions…Sets of questions are more reliable and they give more consistent results". Hence, a set of questions are formulated to better understand not just the contractor's attitude on the prevalence of subcontracting practices, but on the reasons they chose to adopt it.

5.1.3 Sampling Target

The sampling group primarily consists of firms who have main contractor status and subcontractor status. This sampling group was selected, as it is common for contractors, regardless of their status, to subcontract or outsource their work to subcontractors down the hierarchy, which has enabled the existence of the MLSS in Singapore. The need to subcontract is based on work specialisation that has made the MLSS even more complex and prevalent. In addition, the profile of the selected participants is important, as it ensures the accuracy, credibility, and usefulness of the survey's results. Hence, all the contractors that have been interviewed were extracted from the BCA registry database and are of different backgrounds for a wider analysis base. Most importantly, these firms were selected randomly to ensure there is no instance of bias that can compromise the reliability of the results. For details on each individual interview, please refer to Appendix B for the respective interview transcripts.

5.1.4 Objectives and Requirements of Interview Questionnaire

Four main objectives were intended from the series of interviews conducted:

1. To obtain contractors' inputs on the underlying motivations for their preference for adopting subcontracting. Not many publications and previous studies have focused on the reasons that propelled contractors to adopt subcontracting practices in the context of Singapore. Hence, the research team sought to gain insight via interviewing contractors who have done so.
2. To gather the contractor's perspectives on the practice of subcontracting their work, which affects productivity. This determines the extent to which interviewees view subcontracting practices as a means that affect the construction productivity of their firms.
3. To understand the state of the local construction industry. This facilitates understanding for the readers with respect to the purpose of this study.
4. To seek the contractor's recommendations on how to further improve productivity despite the adoption of subcontracting practices. This would set the stage for further studies on subcontracting practices to enable a positive effect in construction productivity in Singapore.

As the interview questionnaire did not require any self-completion, the questionnaire layout was not a concern. Nonetheless, both attitude and factual questions were covered. Several factors mentioned by Oppenheim (1992) were considered when designing the questions posed to interviewees. These are listed as follows:

1. Straightforward questions in a simple sentence structure,
2. Omit leading and biased questions,
3. All questions should be consistent in style,
4. Avoid ambiguous words, proverbs, and double negatives, and
5. Avoid double barrelled questions.

Note: Refer to the interview questionnaire in the Appendix B and C.

5.2 Quantitative Research

The aim of quantitative research is to quantify the data and generalise the results from a sample to the population of interest (Malhotra, 1996). The results produced by the data-mining software will be further analysed by IBM's Statistical Package for Social Sciences (SPSS) to further quantify and generalise the results for increased accuracy and relevance.

5.3 Interview Questionnaire

There are two different types of questionnaires, each with 10 questions. The reason for the different types of questionnaires is due to the presence of two different target groups. In Table 5.1, the intended target group is for 10 main contractors, while in Table 5.2, the other intended target group is the 10 subcontractors. In Tables 5.1 and 5.2, the 10 drafted questions were each categorised into one of the four main objectives that serve to achieve the overall interview objectives. (See Annex 2 and Annex 3 respectively for the questionnaires on main contractors and subcontractors.) These questions serve to guide discussions with company representatives who had vital information on subcontracting practices within their company and the state of the construction industry, which is crucial for the in-depth understanding of subcontracting practices in the local industry and its associated effect on construction productivity.

The profiles of these 20 interviewees are shown in Annex 4 and as tabulated in Table 5.3.

5.4 Data Analysis Methodology

During the interview sessions, the focus was placed on the contractors and the reasons that spurred the adoption of subcontracting by their firms. In addition, it was to determine from their perspective, whether such practices have indeed had an effect on their firm's construction productivity. This is followed by any probable

Table 5.1 Interview questions for main contractors

Contractors' input on the underlying motivations for their preference for adopting subcontracting
Q4. What guiding principles do you adopt to decide on (a) what and (b) how much work to subcontract out to others? Q5. In the midst of a tightening labour market and increases in foreign worker levies which have resulted in a squeeze in foreign worker numbers, do you subcontract more work out to others than ever before? Q7. Do you have a core team of subcontractors whom you always work with? Why?
Contractor's perspectives on the practice of subcontracting their work that affects productivity
Q6. Would subcontracting out your firm's work to others affect your firm's productivity? Q8. How do you help your subcontractors enhance their productivity, if any?
Understanding the state of the local construction industry
Q1. Does your firm subcontract work out to others? Q2. What percentage of work does your firm typically subcontract out to others? Q3. What types of work/trades do you typically subcontract out to others? Q9. Is your firm also a subcontractor?
Contractor's recommendation on how to further improve productivity despite the adoption of subcontracting practices
Q10. If your firm is also a subcontractor, what and how would you want your main contractors to help you level up your firm's productivity?

Table 5.2 Interview questions for subcontractors

Contractors' input on the underlying motivations for their preference for adopting subcontracting
Q4. What guiding principles do you adopt to decide on (a) what and (b) how much work to sub-subcontract out to others? Q5. In the midst of a tightening labour market and increases in foreign worker levies, which have resulted in a squeeze in foreign worker numbers, do you sub-subcontract more work out to others than ever before? Q7. Do you have a core team of sub-subcontractors whom you always work with? Why?
Contractor's perspectives on the practice of subcontracting their works that affects productivity
Q6. Would sub-subcontracting out your firm's work to others affect your productivity? How do you ensure your firm stays productive? Q8. How do you help your sub-subcontractors enhance their firm's productivity, if any?
Understand the state of the local construction industry
Q1. Does your firm sub-subcontract work out to others? Q2. What percentage of work does your firm typically sub-subcontract out to others? Q3. What types of work/trades do you typically sub-subcontract out to others? Q9. How does your firm value-add to your main contractors whom you work with?
Contractor's recommendation on how to further improve productivity despite the adoption of subcontracting practices
Q10. If your firm is a subcontractor, what and how would you want your main contractors to help you level up your firm's productivity?

Table 5.3 Profile of interviewees

Interview session no.	Company role	Date	Firm type	Annex
1	Executive director	16th Jan 2016	MC	5
2	Director	18th Jan 2016	MC	6
3	project director	20th Jan 2016	MC	7
4	Project manager	22nd Jan 2016	MC	8
5	Owner	25th Jan 2016	SC	9
6	Senior contracts manager	26th Jan 2016	MC	10
7	Managing director	28th Jan 2016	MC	11
8	Managing director	27th Jan 2016	MC	12
9	Project manager	30th Jan 2016	MC	13
10	General manager	31st Jan 2016	SC	14
11	Contracts manager	25th Jan 2016	MC	15
12	Manager	2nd Feb 2016	SC	16
13	Senior contracts manager	3rd Feb 2016	MC	17
14	Manager	4th Feb 2016	SC	18
15	Managing director	20th Feb 2016	SC	19
16	Managing director	11th Mar 2016	SC	20
17	Managing director	30th Jan 2016	SC	21
18	Managing director	12th Mar 2016	SC	22
19	Director	9th Apr 2016	SC	23
20	Managing director	15th Apr 2016	SC	24

MC Main Contractor
SC Subcontractor

recommendations that are suggested to improve their firm's productivity in the long run, to the betterment of the local industry. Hence, the questionnaire has already been divided into four main objectives to enable sufficient ground to be covered to better comprehend the nature of subcontracting practices in the local context. More importantly, a qualitative content analysis will be conducted on the interview data to interpret the data and derive crucial findings that answer the hypothesis of this study.

In addition, qualitative research has greatly benefited from theoretical and methodological developments in data analysis during the past few decades (see, e.g., Bernard & Ryan, 1998; Dey, 1993; LeCompte & Schensul, 1999). Usually, such analysis consists of two categories: content and thematic.

5.4.1 *Content Analysis*

Content analysis is defined as a technique for making inferences by systematically and objectively identifying specified characteristics of messages (Holsti, 1969). Moreover, Holsti (1969) suggested that such content analysis can be used for the following:

- Make inferences about the antecedents of communication;
- Describe and make inferences about the characteristics of communication;
- Make inferences about the effects of communication; and
- Analyse subjects in social sciences and humanities.

In addition, content analysis is a research technique that is adopted to systematically make inferences about the attitudes, intentions, and values of individuals by identifying specific characteristics in textual messages (Morris, 1996). As such, Berelson (1971) commented that content analysis is a frequent tool used to analyse recorded transcripts of interviews with participants, especially with conceptual analysis and relational analysis. Furthermore, Morris (1996) listed the advantages of computerised content analysis over human-coded content analysis, such as the following:

- Stability and reliability of the computerised coding scheme and coder (Weber, 1990);
- Explicit coding rules yielding formally comparable results (Weber, 1990);
- Easy manipulation of text to create word-frequency counts, keyword in-context listings, and concordances; and
- Ability to process larger volumes of qualitative data at lower costs (Gephart & Wolfe, 1989).

Moreover, Morris (1996) mentioned that, in the realm of strategic management, the unobtrusive nature of content analysis makes it well suited for such a research study as well. Therefore, the chosen content-analysis software called QDA Miner was utilised in this present study as shown in Fig. 5.2.

A mixed model qualitative data analysis software package capable of coding, annotating, retrieving, and analysing collections of documents or images, it has been lauded by among those shown in Fig. 5.2 as "One of, if not the best multi-use text application", and garnered reviews, such as:

> In sum, QDA Miner possesses outstanding mixed model analytical tools, all of which work well. We are impressed not only by the range of functionality represented in this software, but also by the good design evident in how the program's user interface presents the various components of each tool to the user.

Because of client confidentiality, the above quotes were not attributed to any one users.

With its diverse clientele, its reliability and effectiveness as an analytical tool is widely recognised by corporations and governments worldwide. In addition, for content analysis, the researcher evaluated the saliency or frequency of particular words or phrases within the body of original text data to identify repeated ideas or count the frequency of occurrence. Besides conducting simple word counts, content analysis also includes associated attributes of keywords and other semantic elements, such as synonyms, location in the text, and surrounding phrases or words (Dey, 1993).

More importantly, content-analysis techniques are valued for their reliability and efficiency. With the appropriate software, large quantities of text files can be

Fig. 5.2 QDA miner software and some of its clientele

quickly scanned and keywords can be tallied. As "raw" data is utilised, there is minimal interpretation involved in the word counts, resulting in greater reliability.

5.4.2 Thematic Analysis

On the other hand, thematic analysis is more nuanced and involved. Such analysis focuses on identifying and describing both implicit and explicit ideas, instead of counting explicit words or phrases. Codes developed for themes and ideas are applied or linked to raw data as summary markers for later analysis, which may include comparing the relative frequencies of themes within a data set, finding code co-occurrence, or displaying code relationships via graphical means.

Moreover, both content and thematic analysis can be data-driven, as in the grounded theory (Glaser & Strauss, 1967; Kearney, Murphy, & Rosenbaum, 1994; Wright, 1997), or theory-driven (Krippendorf, 1980; Weber, 1990). In a data-driven approach, the researcher carefully scrutinises the data, looking for keywords, trends, or themes in the data that will help outline the analysis, before any analysis takes place. However, a theory-driven approach is guided by hypotheses or specific ideas that the researcher wants to assess. The researcher may still scrutinise the data prior to analysis, but the researcher's analysis categories have been determined a priori, without consideration of the data. In this study, theory-driven approaches are preferred, as this is more structured and therefore considered more reliable, whereby the same results are likely despite having different coders to conduct the analysis.

This study utilises both content and thematic analysis with a theory-driven approach. With the aid of the software QDA Miner 4.0, which is capable of performing both types of analysis, the 20 interview transcripts will undergo the content-analysis process. To enable better clarity on how such a content-analysis process works, an illustration of the process is shown in Fig. 5.3.

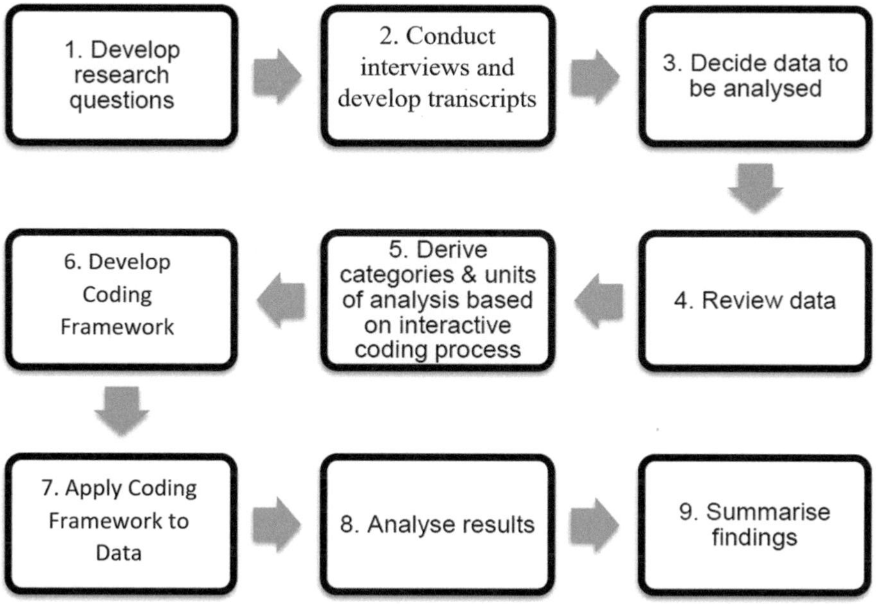

Fig. 5.3 Illustration of content-analysis process

References

Berelson, B. (1971). *Content analysis in communication research*. New York: Free Press.

Bernard, H. R., & Ryan, G. (1998). Text analysis: Qualitative and quantitative methods. In H. R. Bernard (Ed.), *Handbook of methods in cultural anthropology* (pp. 595–645). Walnut Creek, Calif.: Altamira.

Dey, I. (1993). *Qualitative data analysis: A user-friendly guide for social scientists*. New York: Routledge.

Gephart, R. P., & Wolfe, R. A. (1989). Qualitative data analysis: Three microcomputer-supported approaches. In F. Hoy (Ed.), *Academy of management best paper proceedings* (pp. 382–386). Academy of Management, Ada.

Glaser, B., & Strauss, A. (1967). *The discovery of grounded theory: Strategies for qualitative research*. New York: Aldine.

Holsti, O. R. (1969) *Content analysis for the social sciences and humanities*. Addison-Wesley, Reading.

Kearney, M., Murphy, S., & Rosenbaum, M. (1994). Mothering on crack cocaine: A grounded theory analysis. *Social Science and Medicine, 38,* 351–361.

Krippendorf, K. (1980). *Content analysis: An introduction to its methodology*. Beverly Hills, California: Sage.

LeCompte, M. D., & Schensul, J. J. (1999). Analyzing and interpreting ethnographic data. Book five of the ethnographer's toolkit. In: J. J. Schensul, & M. D. LeCompte (Eds.), *Walnut Creek*. California: Altamira Press, a division of Sage Publications.

Malhotra, N. K. (1996). *Marketing research: An applied orientation* (2nd ed.). New Jersey: Prentice-Hall.

Morris, E. (1996). Computerized content analysis in management research: A demonstration of advantages and limitations. *Journal of Management, 20,* 903–931.

Muhammad, F. I., & McGoldrick, P. (2003). *Shopping choices with public transport options—An agenda for the 21st century* (pp. 78–81). Ashgate Publishing Ltd.

Oppenheim, A. N. (1992). *Questionnaire design, interviewing and attitude measurement* (2nd ed., pp. 65–80). New York: Continuum.

Weber, R. P. (1990). *Basic content analysis*. Newbury Park.

Wright, K. B. (1997). Shared ideology in alcoholics anonymous: A grounded theory approach. *Journal of Health Communication, 2*(2), 83–99.

Chapter 6
Research Findings

Abstract This chapter presents the interview findings and the analysis of the results. It explains the reasons for the rampant practice of subcontracting in Singapore through the literature review. This qualitative research study identified the effect on construction productivity due to subcontracting from detailed interviews with both main contractors and subcontractors. Responses from the interviews were analysed using word-mining tools to examine the effect on construction productivity.

This chapter depicts key steps of the content-analysis process and showcases the results of the data collection. Such key steps include the derivation of coding categories and the subsequent formulation of the units of analysis based on the iterative coding process, which is stated in Step 5 of the content-analysis process in Chap. 5. The following subsections will address each step of the content-analysis process that is shown in Fig. 5.3 of Chap. 5.

6.1 Profile of Interviewees

Table 6.1 illustrates the profiles of the 20 interview respondents in terms of designation as well as their corresponding firm types. Some of the interviewees are owners themselves of the construction firms, while others are in senior or mid-tier managerial positions. Senior managers ranged from general managers and executive directors to managing directors, and middle-managers included project managers, senior contract managers, and managerial positions. It is interesting to note that middle-managers from main contractor firms consist of half of the interviews for the main contractors, at five out of 10. As compared to the subcontractor firms, seven out of 10 of the interviewees are owners of the firms. As such, the majority of interviewees who work in subcontractor firms are owners, compared to other interviewees from the main contractor firms. This is typical of Singapore's construction industry hierarchy.

L. Sui Pheng and S. Jia Zheng, *Construction Productivity in the Multilayer Subcontracting System*, Management in the Built Environment,
https://doi.org/10.1007/978-981-13-1831-3_6

Table 6.1 Interviewee profiles

Designations of respondents	Number			Percentage (%)
	Main contractor	Subcontractor	Sub-total	Proportion of designated respondents of total interviews
Owner	2	7	9	45.0
Senior-mgmt.	3	1	4	20.0
Middle-mgmt.	5	2	7	35.0
Total from interviews	10	10	20	100.0

6.1.1 Development of Research Questions

The beginning of the content-analysis process requires the development of specific research questions. The purpose of the interviews is to acquire sufficient knowledge from the respondents, who are current industry practitioners that mostly hold senior roles in main contractor firms or are owners with subcontractor firms, about issues regarding subcontracting practices in the local industry. With reference to both the research objectives of this paper and the purpose of the interviews, the research questions were formulated to enable content analysis. For the purpose of enabling the content analysis, these research questions have been adapted from the interview objectives discussed in Chap. 5 as shown below:

- What is the current state of the local construction industry?
- What are the underlying motivations for their preference for adopting subcontracting?
- What are the views from the contractors with regards to the practice of sub-contracting their work, which affects productivity?
- What would the contractors recommend regarding further improving the industry's productivity despite the rampant practice of subcontracting?

6.1.2 Selection of Relevant Data to Be Analysed

Upon completion of the second step in the content-analysis process, which is mainly to conduct 20 interviews and develop the respective transcripts, the third step is to scrutinise all 20 interview transcripts and exclude irrelevant portions of the interviews that do not contribute to the interview objectives. This is followed by the fourth step, also known as data review (called "data cleaning"), which occurs throughout all 20 transcripts.

Data cleaning is defined as the process of detecting irrelevant and incorrect parts within the transcripts and correcting inaccurate records or data from the transcripts in this context. As such, after the data cleaning is completed, all 20 transcripts will have data that are consistent with other similar data sets. This essentially means that, for instance, within a particular transcript, relevant sections of the transcripts are highlighted and tagged with a word or phrase that best describes the overall section that bears relevance to the interview objectives. This was also done for the remaining transcripts.

6.1.3 Category Derivation and Units of Analysis

With reference to the iterative coding process as shown in Fig. 6.1, common themes were determined. Besides that, common categories are determined after the process of data cleaning. This is to enable the ease of grouping together relevant chunks of data throughout the 20 transcripts, which will aid in the ease of coding later, as this is a crucial part for any content-analysis process. The various common categories are placed in Table 6.2.

6.1.4 Coding Scheme

Once the respective categories were derived, codes were assigned to each category, in accordance to the relevance. This is to facilitate coding of all 20 transcripts in the content-analysis process. Table 6.3 shows the coding scheme adopted.

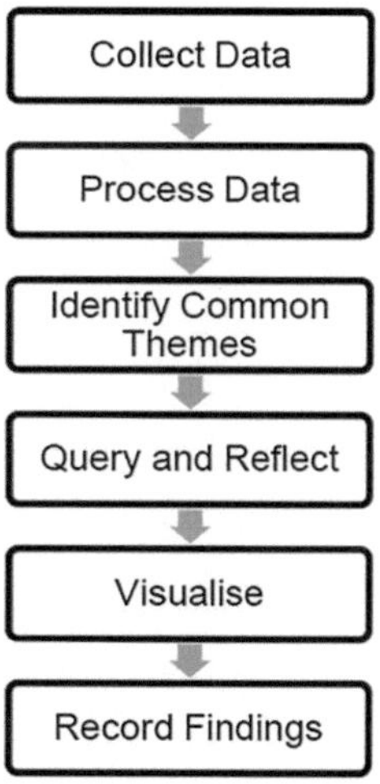

Fig. 6.1 Iterative coding process

Table 6.2 Common categories as units of analysis

S/N	Category	Definition	Coding rules
1	Sub-contractors	Construction firms that enter into a contract or several contracts with a main contractor to perform some of the main contractor's workload. Usually have low construction turnover. Small workforce but conducts specialised construction work	The words within the transcript need to refer to the category. They need not be exact phrase or word but should have similar meaning
2	Main contractors	Construction firms that successfully tendered for a project from the client/owner and liaise directly with them to meet their project needs. Usually have high construction turnover. Large workforce but conducts mainly general work	
3	External forces	Defined as forces beyond the control of any firms that have a significant effect on the firms' operations	
4	Flexibility	Defined as greater use of an adjustable periphery, less rigid work practices, and implementation and utilisation of new technology that may contribute to a firm's profitability and competitiveness (Hunter, McGregor, Macinnes, & Sproull, 1993)	
5	Multiple tiers	Presence of a hierarchy-based subcontracting system when a subcontractor is involved in multiple projects and further subcontracts work to other subcontractors (Ofori & Lim, 2009)	
6	Perceived benefits	The perception of the positive consequences that are caused by a specific action/actions	
7	Outcomes	A result or consequence due to something that has occurred	

Table 6.3 Coding scheme

S/N	Category	Code
1	Subcontractors	Small Employment Size
		Low Construction Turnover/Earnings
		Specialised work
		Small Scale Work
2	Main contractors	Large Employment Size
		High Construction Turnover
		Undertake General Work
		Large-scale Work
3	External forces	Economic Downturn
		Change in Government Policies
		Globalisation
		Industry Irregularities
4	Flexibility	Work Arrangement
		Resource Deployment
		Enables Resource Control
		Resource Sharing
5	Multiple tiers	Subcontract/Outsourcing by SC to SC
		Subcontract/Outsourcing by MC to SC
6	Perceived benefits	Efficient
		Achieves SCA
		Accrue Firm Flexibility
7	Outcomes	Reduced Resource Wastage
		Increased Resource Wastage
		Improved Work Productivity
		Decreased Work Productivity

6.1.5 Coding of Data

Based on the research questions for this content-analysis process, a proposed coding analytical framework is derived as shown in Fig. 6.2. The framework serves to aid in the coding of data by giving a clearer representation of the relationships among the code categories and their respective codes and relative positions in this overarching theme of subcontracting practices in Singapore's construction industry. As such, any phrases, sentences, or words that bear similar meaning in the 20 transcripts to the code Resource Deployment are tagged to this code during the coding stages of the content analysis. This coding sequence is the same for all other codes as well.

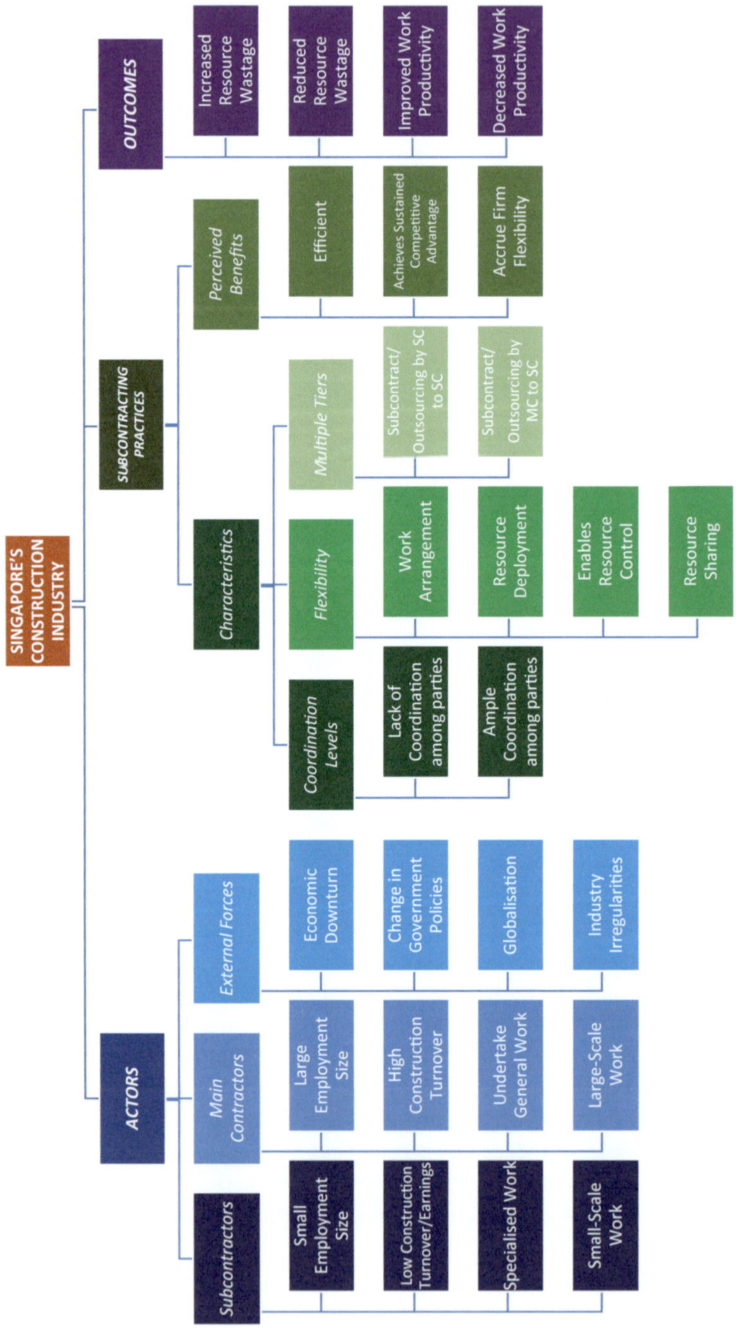

Fig. 6.2 Proposed coding analytical framework

6.1.6 Apply Coding Scheme to Data

Steps to programme codes using QDA Miner

The following steps showcase how the coding scheme shown in Fig. 6.2 is applied to the 20 interview transcripts that have been uploaded into the software.

Step 1: Select option "Create a new project" as shown in Fig. 6.3.
Step 2: Input 20 transcripts into the software as shown in Fig. 6.4.
Step 3: Input proposed coding analytical framework into software for content coding.

The end result should be depicted within the box located on the bottom left-hand corner that is titled 'CODES'. The content in the box should have a framework that resembles the proposed coding analytical framework as shown in Fig. 6.5. This is also known as the Codebook in the software.

Step 4: Code text segments corresponding to their respective assigned codes.

These text segments have been predetermined prior to the commencement of this software coding process. This was done by reading through all 20 interview transcripts and classifying which text segments should be categorised into their respective assigned codes via marking the relevant parts of the interview transcripts. Note that the software allows the user to select the colour of the code (blue) that is associated with the selected text for the case, as shown in Fig. 6.6.

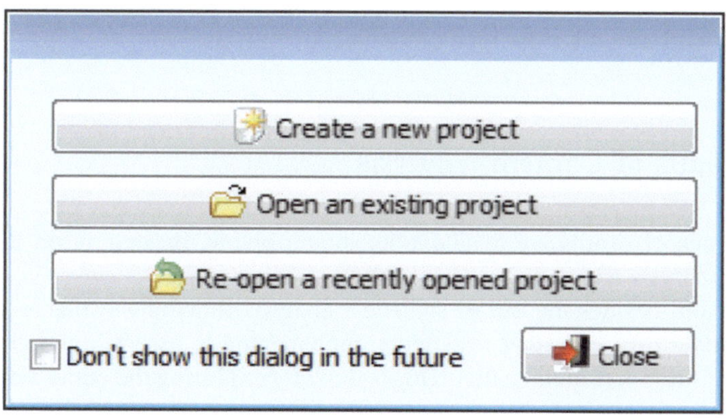

Fig. 6.3 Screenshot to "create a new project" in QDA miner

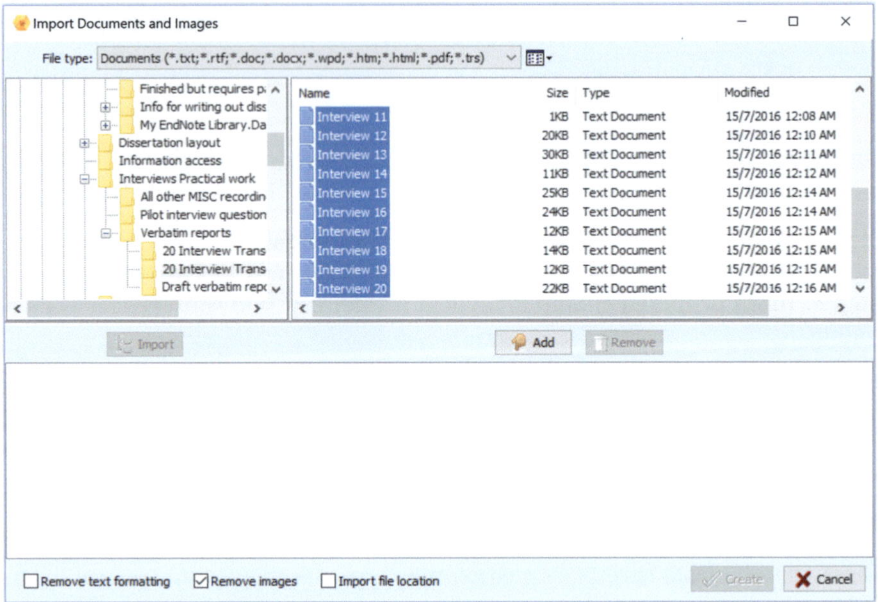

Fig. 6.4 Screenshot of the input process of 20 transcripts

Step 5: Once coding of text segments is completed, the software user will have a screen view that is similar to Fig. 6.7.

The user can then proceed to analyse the content using the software function *analyse* to conduct content analysis on the 20 transcripts. Various types of results will be derived based on the type of statistical analysis.

6.2 Results of Content Analysis

Various types of results are available as shown below, depending on the type of statistical analysis adopted.

In addition, besides the use of statistical analysis functions available within the QDA Miner software, further tests were implemented to ascertain the validity and accuracy of the correlation results. This is done by exporting the current data format of the results from QDA Miner in the file format for Microsoft Excel and then running further correlation analyses in IBM's SPSS. A screenshot of the export of the data format of the results from QDA Miner is as shown in Fig. 6.18.

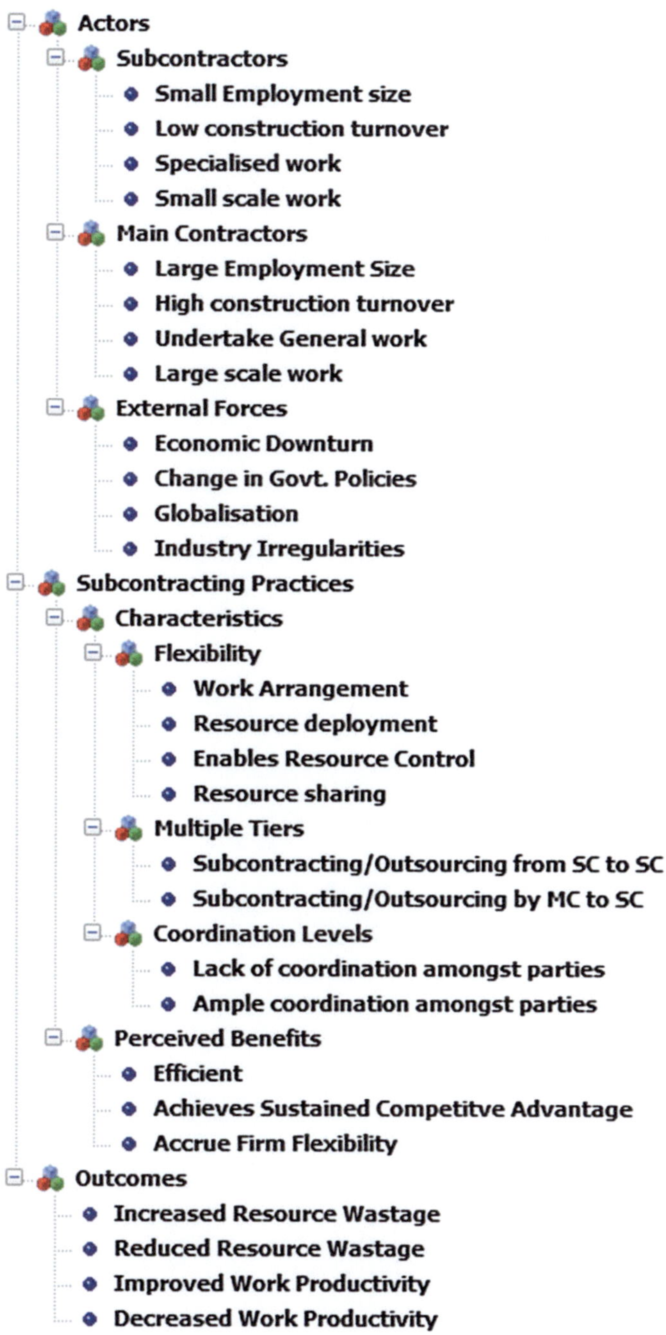

Fig. 6.5 Input 'CODES' for content coding and analysis

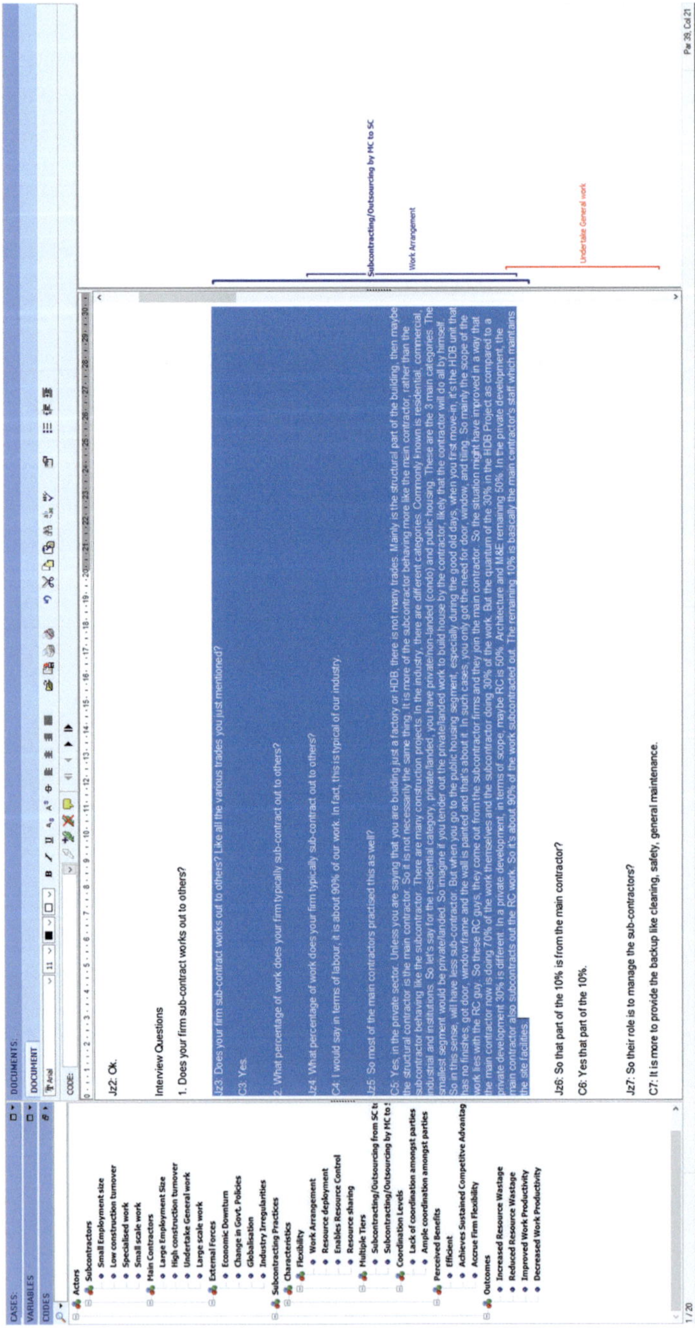

Fig. 6.6 Screenshot of coding the text segments within transcripts

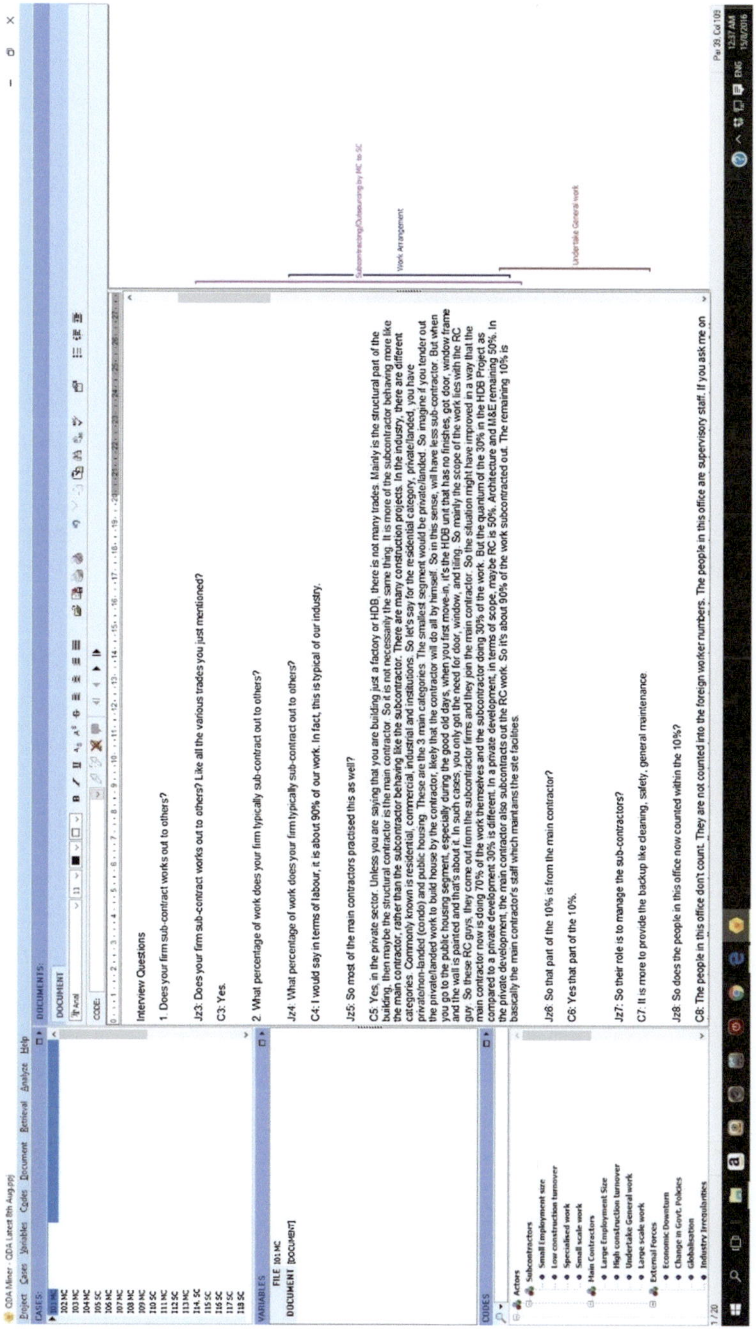

Fig. 6.7 Screenshot after completion of coding process

Fig. 6.8 Word plot results

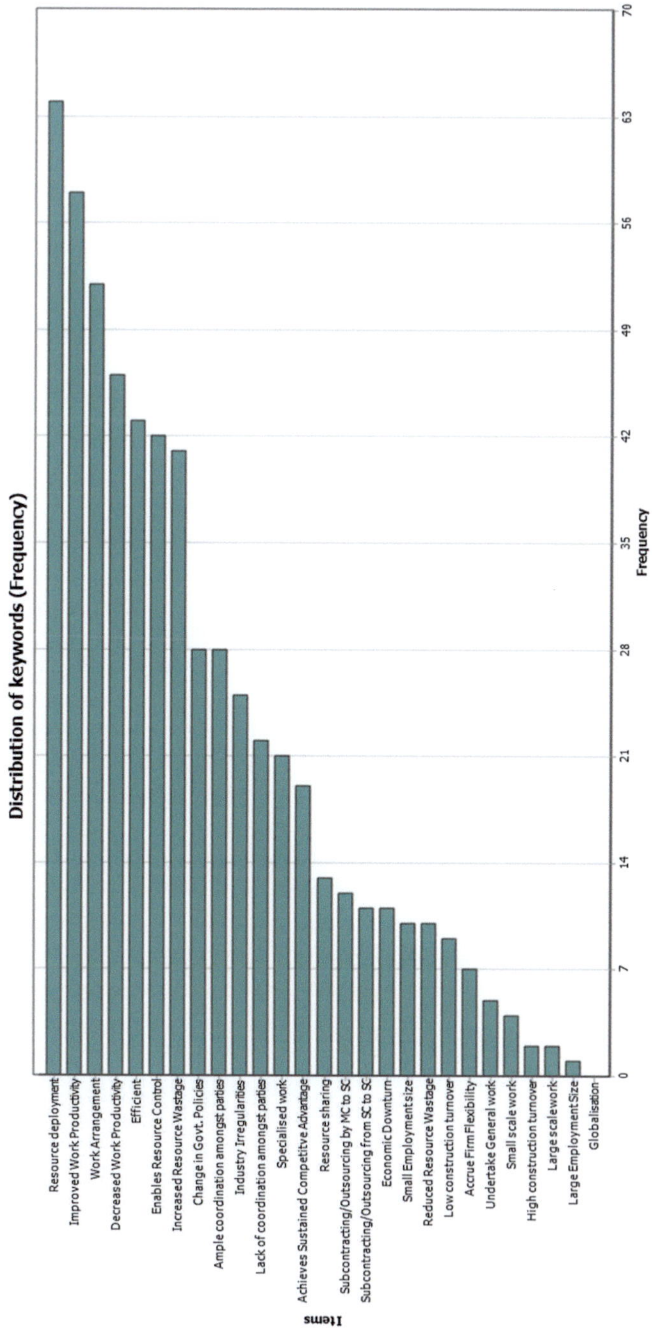

Fig. 6.9 Bar chart results of code frequency distribution

All nine types of results will be discussed and analysed in Chap. 7, coupled with appropriate quotes from various transcripts to support the results.

1. *Results* via *word plot* (Fig. 6.8)
2. *Results* via *distribution of code frequency (bar chart)* (Fig. 6.9)
3. *Results* via *distribution of code frequency (pie chart)* (Fig. 6.10)
4. *Results of coding co-occurrences (Jaccard's coefficient of occurrence, similarity index)* (Figs. 6.11, 6.12, and 6.13)
5. *Results of coding co-occurrences (Jaccard's coefficient of occurrence, agglomeration order)* (Fig. 6.14)
6. *Frequency of coding co-occurrences in all 20 transcripts* (Fig. 6.15)
7. *Frequency matrix of codes in all 20 transcripts* (Fig. 6.16)
8. *Proximity results* via *Jaccard's Coefficient* (Figs. 6.17 and 6.18)
9. *Correlation analysis* via *SPSS* (Fig. 6.19).

6.3 Validity of Analysis

Although the approach of employing qualitative analysis is uncommon for such a study, it nonetheless shed new light on various significant factors that contribute to the prevalent practice of subcontracting, the effects of such practices on various players in the industry, and most importantly, its effect on construction productivity. To reiterate, common coding categories were derived from reviewing all 20 interview transcripts and were put into Table 6.2 for clearer representation. These coding categories were not derived from existing literature reviews so as not to limit the scope of this study, while enabling an in-depth investigation into the research problem.

To ensure further accuracy and credibility of the study results, the data collection and analytical process were both methodologically drafted and controlled. Interviewees were rigorously selected; not only must they have had many years of experience, they ought to have a good grasp of knowledge in the industry. Most importantly, their firm must also be of good standing and reputation in the local market.

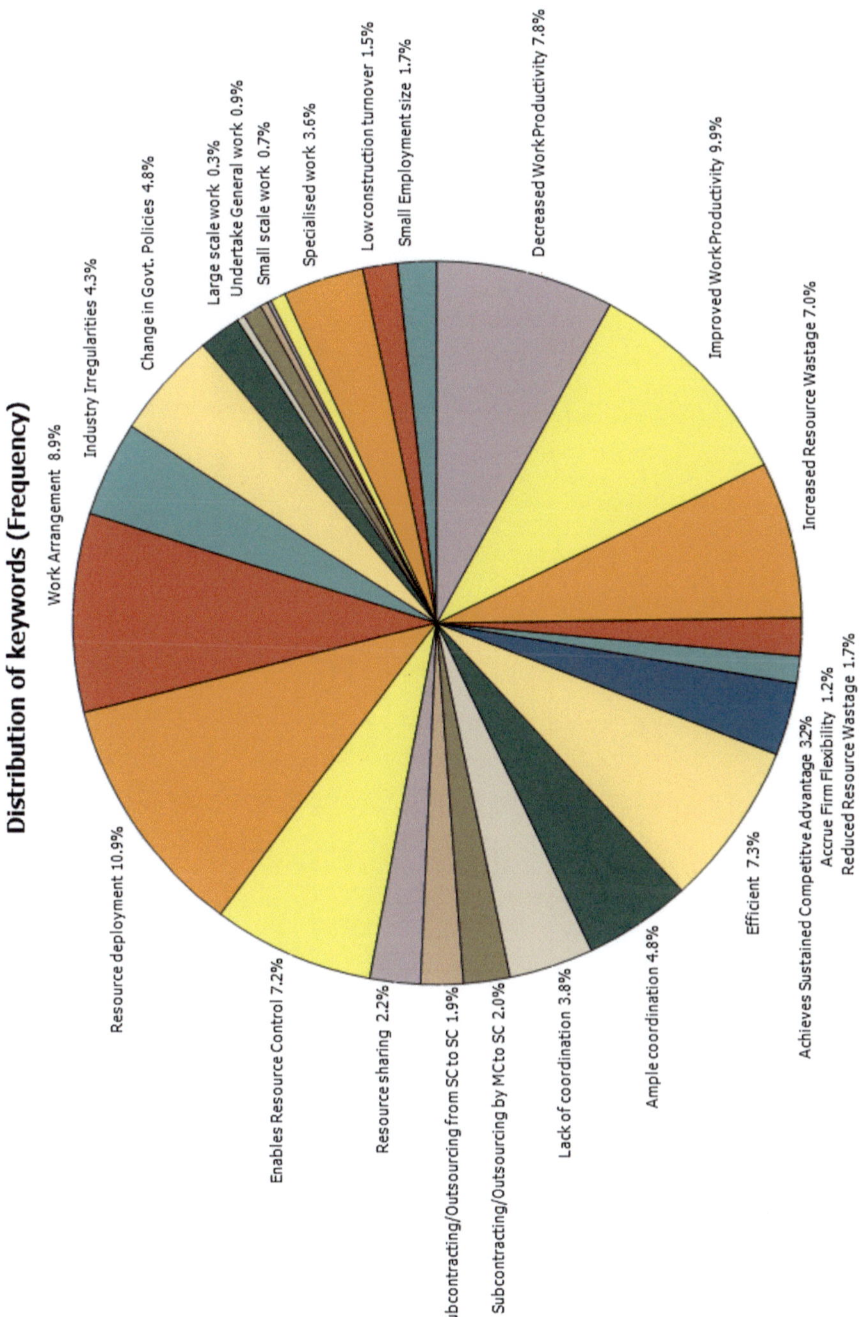

Fig. 6.10 Pie chart results of code frequency distribution

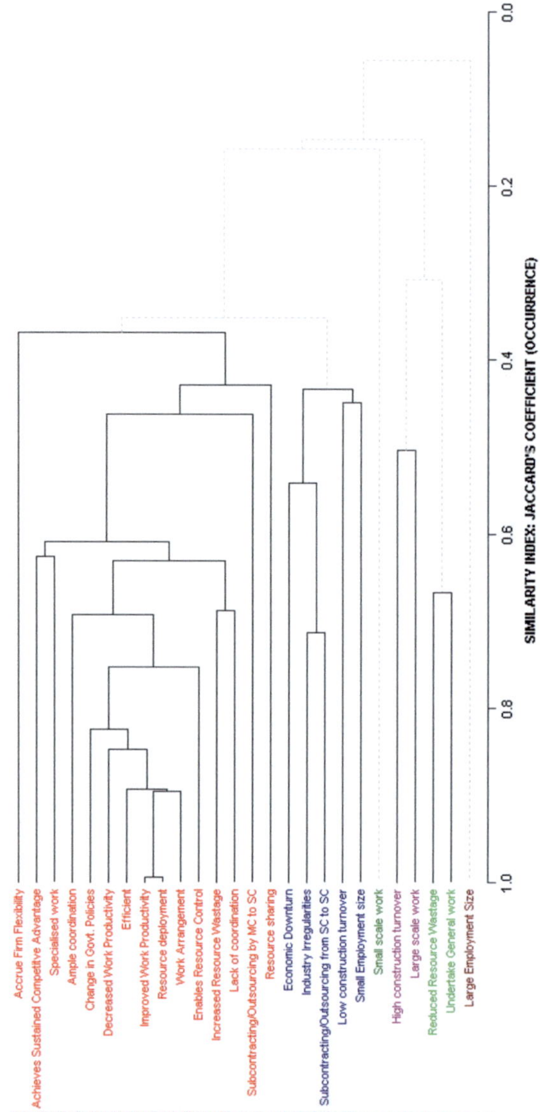

Fig. 6.11 Results of coding co-occurrences

Node	Group 1	Group 2	Similarity
1	Improved Work Productivity	Resource deployment	1.000
2	Node 1	Work Arrangement	0.900
3	Efficient	Node 2	0.898
4	Decreased Work Productivity	Node 3	0.850
5	Change in Govt. Policies	Node 4	0.827
6	Node 5	Enables Resource Control	0.754
7	Industry Irregularities	ıbcontracting/Outsourcing from SC to	0.714
8	Ample coordination	Node 6	0.693
9	Increased Resource Wastage	Lack of coordination	0.688
10	Reduced Resource Wastage	Undertake General work	0.667
11	Node 8	Node 9	0.630
12	chieves Sustained Competitve Advanta	Specialised work	0.625
13	Node 12	Node 11	0.607
14	Economic Downturn	Node 7	0.538
15	High construction turnover	Large scale work	0.500
16	Node 13	;ubcontracting/Outsourcing by MC to S	0.458
17	Low construction turnover	Small Employment size	0.444
18	Node 14	Node 17	0.428
19	Node 16	Resource sharing	0.423
20	Accrue Firm Flexibility	Node 19	0.361
21	Node 20	Node 18	0.344
22	Node 15	Node 10	0.300
23	Node 21	Small scale work	0.147
24	Node 23	Node 22	0.136
25	Node 24	Large Employment Size	0.043

Fig. 6.12 Table of similarity index

	Accrue Firm Flexibility	Achieves Sustained Competitve Advantage	Ample coordination	Change in Govt. Policies	Decreased Work Productivity	Economic Downturn	Efficient	Enables Resource Control	High construction turnover	Improved Work Productivity	Increased Resource Wastage	Industry Irregularities	Lack of coordination	Large Employment Size	Large scale work	Low construction turnover	Reduced Resource Wastage	Resource deployment	Resource sharing	Small Employment size	Small scale work	Specialised work	Subcontracting/Outsourcing by MC to SC	Subcontracting/Outsourcing from SC to SC	Undertake General work	Work Arrangement
Accrue Firm Flexibility	1																									
chieves Sustained Competitve Advantage	0.429	1																								
Ample coordination	0.222	0.556	1																							
Change in Govt. Policies	0.333	0.579	0.684	1																						
Decreased Work Productivity	0.389	0.632	0.65	0.842	1																					
Economic Downturn	0.071	0.313	0.438	0.316	0.368	1																				
Efficient	0.389	0.632	0.65	0.75	0.8	0.3	1																			
Enables Resource Control	0.278	0.611	0.632	0.737	0.789	0.5	0.7	1																		
High construction turnover	0	0.154	0.133	0.118	0.053	0.111	0.111	0.125	1																	
Improved Work Productivity	0.35	0.65	0.75	0.85	0.9	0.4	0.9	0.8	0.1	1																
Increased Resource Wastage	0.4	0.588	0.611	0.632	0.778	0.294	0.6	0.579	0.067	0.7	1															
Industry Irregularities	0.111	0.368	0.556	0.579	0.632	0.615	0.55	0.611	0.071	0.65	0.5	1														
Lack of coordination	0.538	0.625	0.4	0.579	0.722	0.105	0.722	0.526	0.071	0.65	0.688	0.368	1													
Large Employment Size	0.143	0.077	0	0.059	0.056	0	0.056	0.063	0	0.05	0	0	0.077	1												
Large scale work	0	0.077	0.067	0.059	0	0.125	0.056	0.063	0.5	0.05	0	0.077	0	0	1											
Low construction turnover	0.083	0.188	0.167	0.278	0.333	0.273	0.263	0.222	0	0.3	0.25	0.462	0.267	0	0	1										
Reduced Resource Wastage	0.2	0.385	0.333	0.294	0.211	0.083	0.278	0.235	0.4	0.25	0.267	0.125	0.286	0	0.2	0	1									
Resource deployment	0.35	0.65	0.75	0.85	0.9	0.4	0.9	0.8	0.1	1	0.7	0.65	0.65	0.05	0.05	0.3	0.25	1								
Resource sharing	0.143	0.467	0.333	0.444	0.5	0.214	0.421	0.471	0.1	0.45	0.438	0.375	0.467	0.111	0	0.364	0.167	0.45	1							
Small Employment size	0.167	0.333	0.222	0.263	0.389	0.25	0.316	0.278	0	0.35	0.4	0.538	0.429	0	0	0.444	0.091	0.35	0.231	1						
Small scale work	0	0.067	0.125	0.176	0.167	0	0.167	0.118	0	0.15	0.133	0.231	0.231	0	0	0.286	0	0.15	0.2	0.25	1					
Specialised work	0.429	0.625	0.556	0.579	0.632	0.4	0.632	0.611	0.071	0.65	0.588	0.444	0.529	0.077	0.077	0.188	0.286	0.65	0.375	0.25	0.067	1				
Subcontracting/Outsourcing by MC to SC	0.417	0.533	0.471	0.5	0.4	0.125	0.474	0.444	0.2	0.5	0.412	0.15	0.438	0.1	0.1	0	0.364	0.5	0.267	0.063	0.083	0.353	1			
ubcontracting/Outsourcing from SC to SC	0.2	0.333	0.368	0.474	0.611	0.462	0.526	0.5	0	0.55	0.389	0.714	0.412	0.091	0	0.545	0.067	0.55	0.429	0.5	0.167	0.5	0.05	1		
Undertake General work	0.2	0.286	0.333	0.294	0.211	0.083	0.278	0.235	0.4	0.25	0.267	0.125	0.286	0	0.2	0	0.667	0.25	0.167	0	0.143	0.2	0.5	0	1	
Work Arrangement	0.389	0.55	0.737	0.842	0.8	0.3	0.895	0.7	0.111	0.9	0.6	0.55	0.632	0.056	0.056	0.263	0.278	0.9	0.421	0.25	0.167	0.632	0.474	0.526	0.278	1

Fig. 6.13 Table of similarity coding co-occurrences

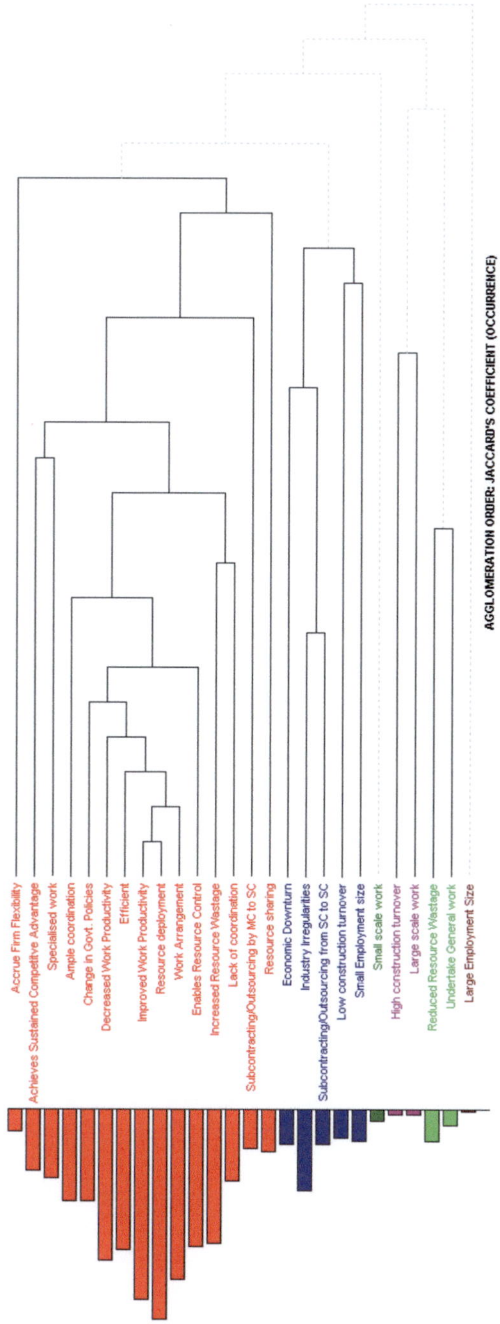

Fig. 6.14 Coding co-occurrences results in agglomeration order

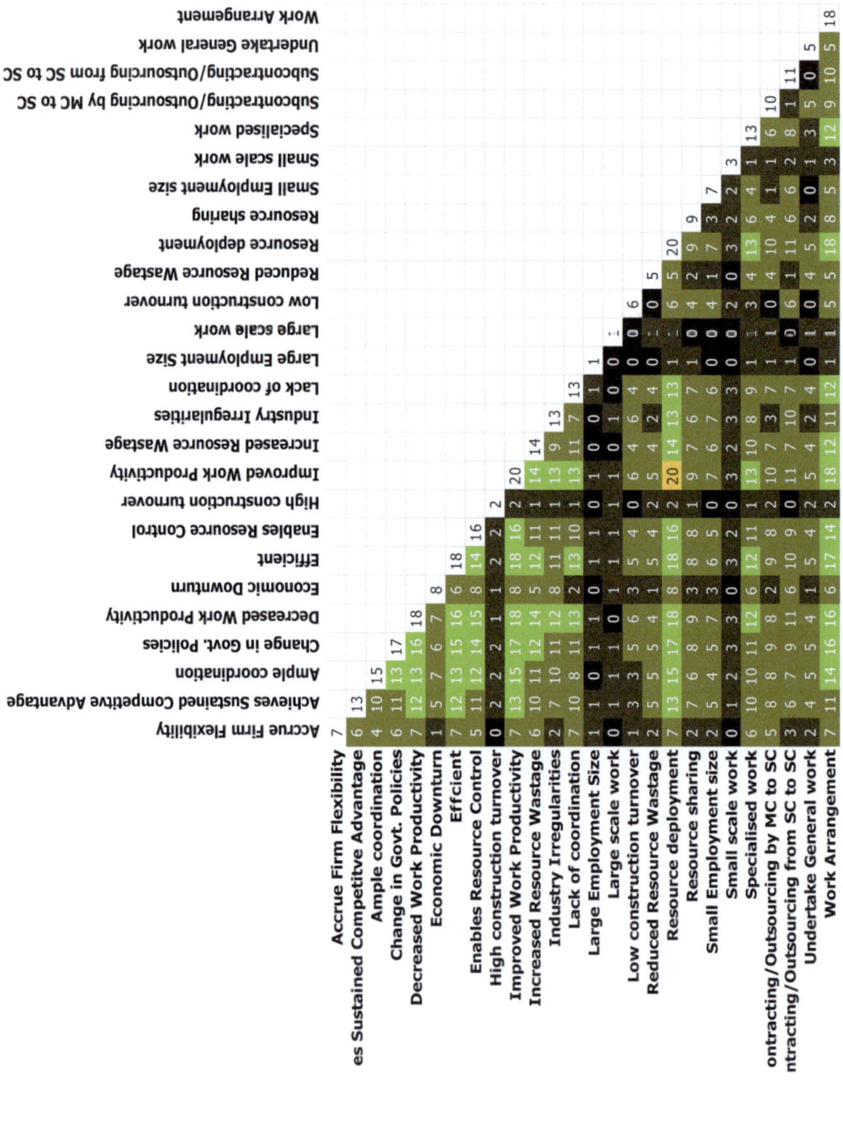

Fig. 6.15 Coding co-occurrences frequency results in 20 transcripts

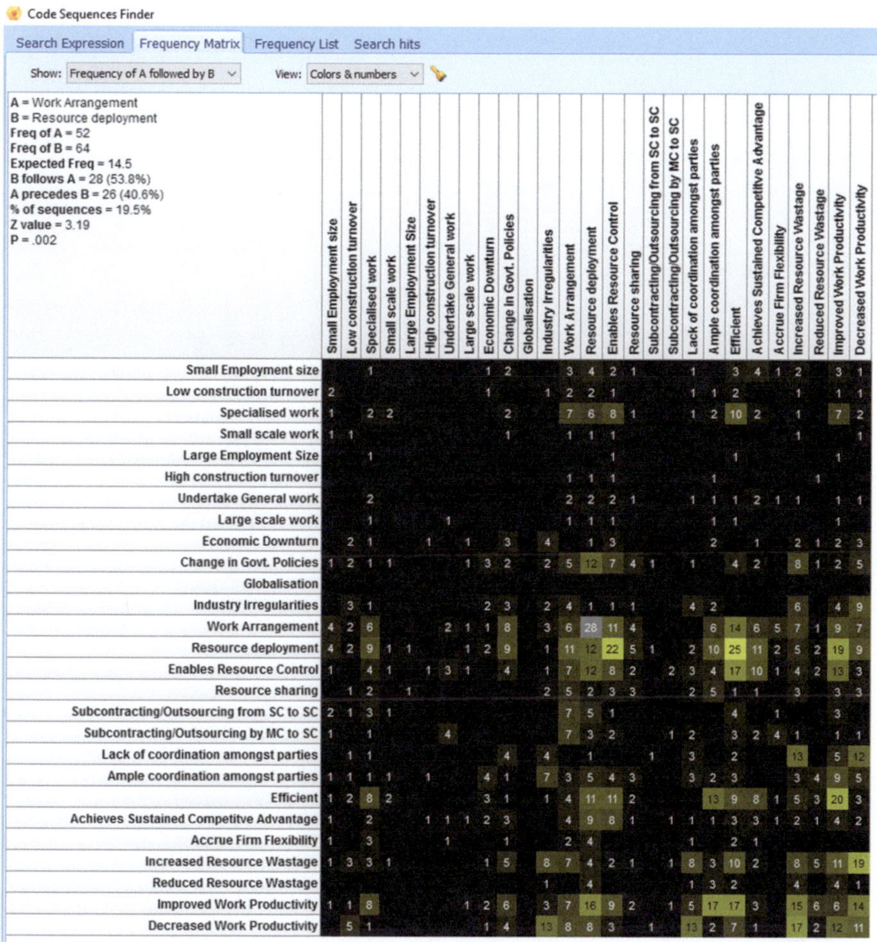

Fig. 6.16 Results of frequency matrix from 20 transcripts

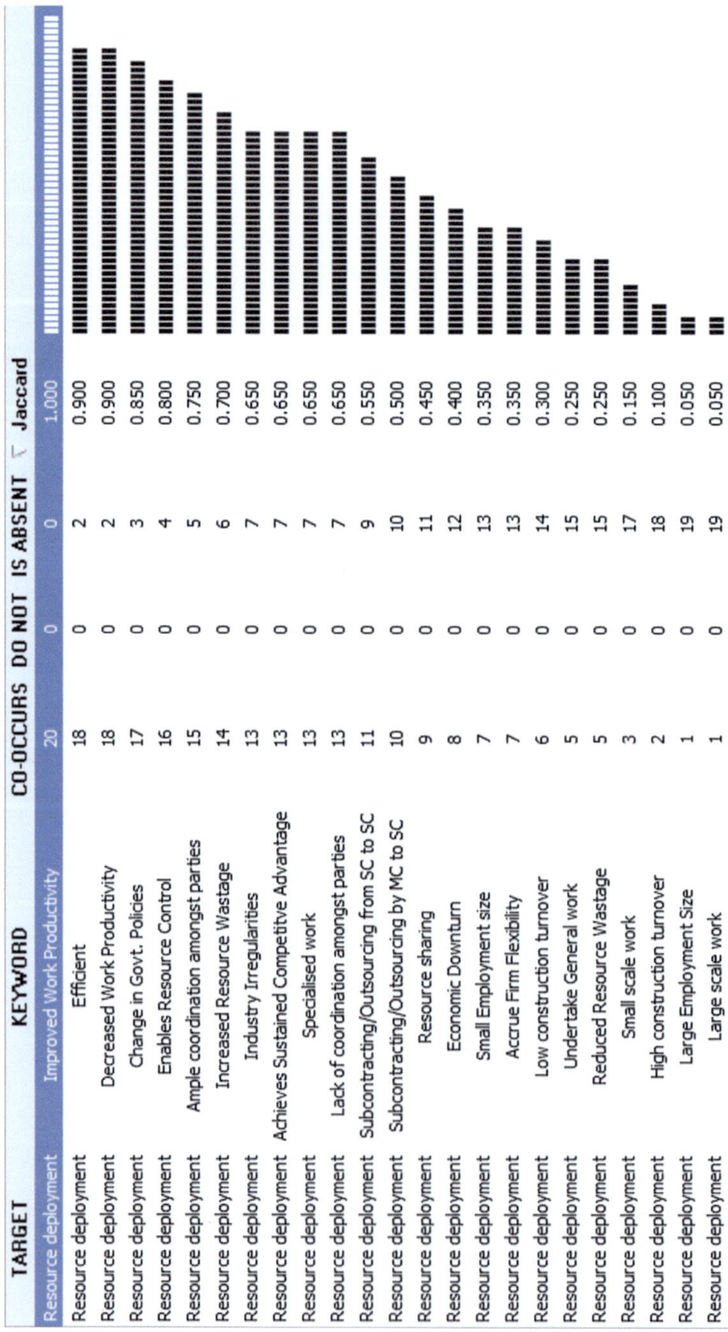

TARGET	KEYWORD	CO-OCCURS	DO NOT	IS ABSENT	Jaccard
Resource deployment	Improved Work Productivity	20	0	0	1.000
Resource deployment	Efficient	18	0	2	0.900
Resource deployment	Decreased Work Productivity	18	0	2	0.900
Resource deployment	Change in Govt. Policies	17	0	3	0.850
Resource deployment	Enables Resource Control	16	0	4	0.800
Resource deployment	Ample coordination amongst parties	15	0	5	0.750
Resource deployment	Increased Resource Wastage	14	0	6	0.700
Resource deployment	Industry Irregularities	13	0	7	0.650
Resource deployment	Achieves Sustained Competitive Advantage	13	0	7	0.650
Resource deployment	Specialised work	13	0	7	0.650
Resource deployment	Lack of coordination amongst parties	13	0	7	0.650
Resource deployment	Subcontracting/Outsourcing from SC to SC	11	0	9	0.550
Resource deployment	Subcontracting/Outsourcing by MC to SC	10	0	10	0.500
Resource deployment	Resource sharing	9	0	11	0.450
Resource deployment	Economic Downturn	8	0	12	0.400
Resource deployment	Small Employment size	7	0	13	0.350
Resource deployment	Accrue Firm Flexibility	7	0	13	0.350
Resource deployment	Low construction turnover	6	0	14	0.300
Resource deployment	Undertake General work	5	0	15	0.250
Resource deployment	Reduced Resource Wastage	5	0	15	0.250
Resource deployment	Small scale work	3	0	17	0.150
Resource deployment	High construction turnover	2	0	18	0.100
Resource deployment	Large Employment Size	1	0	19	0.050
Resource deployment	Large scale work	1	0	19	0.050

Fig. 6.17 Proximity results

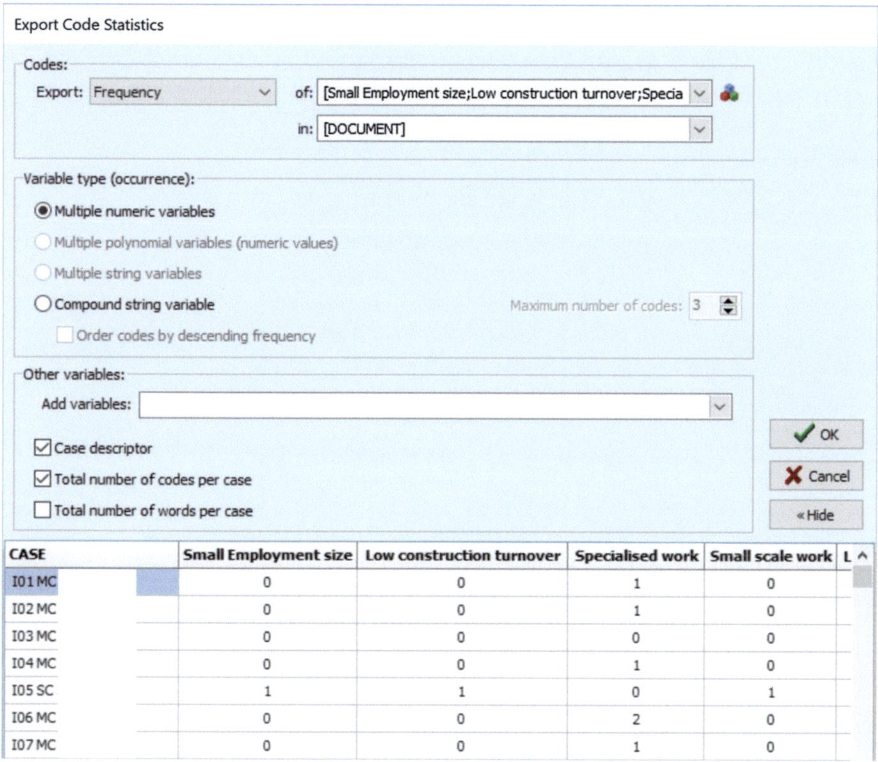

Fig. 6.18 Screenshot of the export of data format results from QDA miner

Correlations

		Small Employment size	Low construction turnover	Specialised work	Small scale work	Large Employment Size	High construction turnover	Undertake General work	Large scale work	Economic Downturn	Change in Govt. Policies	Globalisation	Industry Irregularities	Work Arrangement	Resource deployment	Enables Resource Control	Resource sharing	Subcontracting /Outsourcing from SC to SC	Subcontracting /Outsourcing by MC to SC	Lack of coordination	Ample coordination	Efficient	Achieves Sustained Competive Advantage	Accrue Firm Flexibility	Reduced Resource Wastage	Increased Resource Wastage	Improved Work Productivity	Decreased Work Productivity
Change in Govt. Policies.	Pearson Correlation	0.078	0.346	-0.076	0.160	0.160	0.039	-0.134	-0.107	0.047	1	.[a]	0.118	.593*	0.334	0.034	0.327	0.3[illegible]4	-0.245	-0.212	-0.010	-0.224	-0.164	0.024	-0.126	0.080	-0.086	0.159
	Sig. (2-tailed)	0.743	0.136	0.750	0.502	0.500	0.871	0.573	0.654	0.844			0.621	0.006	0.150	0.887	0.159	0.1[illegible]3	0.297	0.371	0.966	0.343	0.489	0.919	0.596	0.737	0.717	0.502
	N	20	20	20	20	20	20	20	20	20	20	20	20	20	20	20	20	[illegible]0	20	20	20	20	20	20	20	20	20	20
Decreased Work Productivity	Pearson Correlation	0.402	0.133	-0.231	0.339	-0.041	-0.259	0.103	-0.315	-.456*	0.159	.[a]	0.292	0.149	0.088	-0.111	0.248	0.2[illegible]2	-0.117	.697*	-0.397	0.157	-0.217	0.056	0.291	.740*	0.032	1
	Sig. (2-tailed)	0.079	0.576	0.327	0.143	0.864	0.271	0.665	0.176	0.044	0.502		0.211	0.531	0.712	0.642	0.291	0.229	0.624	0.001	0.083	0.510	0.358	0.814	0.213	0.000	0.893	
	N	20	20	20	20	20	20	20	20	20	20	20	20	20	20	20	20	20	20	20	20	20	20	20	20	20	20	20
Efficient	Pearson Correlation	-0.125	-0.167	0.237	0.029	-0.025	0.210	0.363	.484*	0.117	-0.224	.[a]	-0.044	0.083	.500*	0.103	-0.215	0.02[illegible]	0.011	0.096	0.281	1	0.167	0.151	0.140	0.115	.507*	0.157
	Sig. (2-tailed)	0.600	0.480	0.315	0.903	0.915	0.375	0.116	0.031	0.622	0.343		0.855	0.729	0.025	0.666	0.363	0.913	0.963	0.688	0.230		0.482	0.525	0.555	0.630	0.022	0.510
	N	20	20	20	20	20	20	20	20	20	20	20	20	20	20	20	20	2[illegible]	20	20	20	20	20	20	20	20	20	20
Improved Work Productivity	Pearson Correlation	-0.137	-0.279	0.350	-0.040	0.171	.473*	0.273	.481*	0.050	-0.086	.[a]	-0.057	0.093	.483*	0.399	-0.226	0.00[illegible]	-0.092	0.039	.461*	.507*	.473*	-0.092	0.366	-0.034	1	0.032
	Sig. (2-tailed)	0.566	0.234	0.130	0.868	0.472	0.035	0.244	0.032	0.833	0.717		0.811	0.695	0.031	0.081	0.338	0.97[illegible]	0.701	0.871	0.041	0.022	0.035	0.699	0.112	0.887		0.893
	N	20	20	20	20	20	20	20	20	20	20	20	20	20	20	20	20	2[illegible]	20	20	20	20	20	20	20	20	20	20
Resource deployment	Pearson Correlation	0.157	0.115	0.174	0.015	0.142	0.336	0.179	.499*	0.147	0.334	.[a]	-0.079	.549*	1	0.372	-0.255	0.14[illegible]	-0.199	-0.126	0.080	.500*	0.346	0.212	0.126	-0.107	.483*	0.088
	Sig. (2-tailed)	0.509	0.628	0.463	0.949	0.549	0.147	0.450	0.025	0.537	0.150		0.741	0.012		0.107	0.278	0.55[illegible]	0.400	0.595	0.736	0.025	0.135	0.371	0.596	0.653	0.031	0.712
	N	20	20	20	20	20	20	20	20	20	20	20	20	20	20	20	20	2[illegible]	20	20	20	20	20	20	20	20	20	20
Work Arrangement	Pearson Correlation	0.000	0.345	0.208	0.240	-0.195	0.425	0.184	0.293	0.158	.593*	.[a]	0.287	1	.549*	.452*	0.349	0.12[illegible]	-0.248	-0.031	0.188	0.083	0.133	-0.123	0.115	0.076	0.093	0.149
	Sig. (2-tailed)	1.000	0.137	0.378	0.309	0.410	0.062	0.437	0.211	0.506	0.006		0.219		0.012	0.046	0.132	0.59[illegible]	0.291	0.898	0.427	0.729	0.577	0.607	0.628	0.750	0.695	0.531
	N	20	20	20	20	20	20	20	20	20	20	20	20	20	20	20	20	2[illegible]	20	20	20	20	20	20	20	20	20	20
Enables Resource Control	Pearson Correlation	-0.263	-0.129	0.135	-0.067	-0.011	.798*	0.423	0.437	0.162	0.034	.[a]	-0.091	.452*	0.372	1	-0.009	-0.201	0.066	-0.028	0.279	0.103	.560*	-0.189	0.238	-0.105	0.399	-0.111
	Sig. (2-tailed)	0.262	0.589	0.572	0.779	0.963	0.000	0.063	0.054	0.496	0.887		0.703	0.046	0.107		0.971	0.395	0.781	0.906	0.234	0.666	0.010	0.424	0.311	0.659	0.081	0.642
	N	20	20	20	20	20	20	20	20	20	20	20	20	20	20	20	20	20	20	20	20	20	20	20	20	20	20	20

*. Correlation is significant at the 0.05 level (2-tailed).

a. Cannot be computed because at least one of the variables is constant.

Fig. 6.19 Correlation results from SPSS

References

Hunter, L., McGregor, A., Macinnes, J., & Sproull, A. (1993). The 'flexible firm': Strategy and segmentation. *British Journal of Industrial Relations, 31*(3), 383–407.

Ofori, G., & Lim, K. H. (2009). Multi-layered subcontracting system in Singapore's construction industry. In K. Kobayashi, A. R. Kharuddin, G. Ofori, & S. Ogunlana (Eds.), *Joint ventures in construction* (pp. 216–226). Great Britain: Thomas Telford Limited.

Chapter 7
Analysis of Research Findings

Abstract This chapter presents the findings from the fieldwork analysis in relation to construction productivity. It highlights the industry sentiments that subcontracting practices have a significant effect on construction productivity. It discusses the research findings from qualitative data obtained from 20 in-depth interviews. This chapter generates and discusses the statistical results using the QDA Miner 4.0 software to conduct data mining on all interview transcripts.

This chapter will discuss the research findings that have been presented in Chap. 6, which consists of data obtained from the 20 interviews. Subsequently, a software called QDA Miner 4.0 was employed to conduct data mining on all interview transcripts that were input into the software. The statistical results from QDA Miner 4.0 were generated, and its findings were discussed.

More importantly, a distinction must be made at this point to ensure clarity of analysis. The commonly used term 'codes', mentioned in Chaps. 5 and 6, will be referred to as 'factors' from this point onwards. Although these two terms generally refer to the same thing in this study, they are used under different contexts. The term 'codes' is used in the context of referring to the factors that are input into the coding stage of the software, whereas the term 'factors' is used during the analysis phase to explain the relationship and significance of one or several 'factors' with regards to the reason for the prevalent adoption of subcontracting practices and how they ultimately affect construction productivity in Singapore.

Table 7.1 shows several content-analysis techniques or methods to extract and interpret information from the 20 transcripts. These are selected methods that would be applied in the study's collection of interview transcripts, so that comparisons can be made and analyses can be done from a few perspectives. It also includes the use of graphs generated by the content-analysis software to enable better visual clarity of the results. Lastly, correlation analysis further determines the strength of the relationships between the different factors and its significance in determining the key reasons for the prevalent adoption of subcontracting practices and their associated effects on Singapore's construction productivity.

L. Sui Pheng and S. Jia Zheng, *Construction Productivity in the Multilayer Subcontracting System*, Management in the Built Environment,
https://doi.org/10.1007/978-981-13-1831-3_7

Table 7.1 Methods of content analysis

Method of content analysis	Premise/basis of measurement
Frequency	Frequency
Sequence matrix	Sequence
Co-occurrences	Co-occurrence
Proximity	Jaccard's coefficient $J(A,B) = \frac{\lvert A \cap B \rvert}{\lvert A \cup B \rvert}$
Similarity	Similarity of a code's co-occurrence patterns with all the other codes
Similarity dendrogram	Similarity of a code's co-occurrence patterns with all the other codes
Strength of links between codes	Proximity

7.1 Frequency of Factor Occurrences

The following results show the frequency of factor occurrences, presented in various statistical forms, such as word clouds, tables, bar charts, and pie charts.

Figure 7.1 is a unique statistical result called a word cloud. A word cloud is a form of data visualisation method, where the more often a particular word appears in a source of textual data, the bigger and bolder it appears in the word cloud (Boost Labs, 2014). It is also defined as an image composed of words used in texts, in which the size of each word indicates its frequency or importance to the given text (Oxford Dictionary, 2016).

As observed, the largest word clouds are the words Resource Deployment in orange and Improved Work Productivity in yellow. In the context of QDA Miner 4.0 software, this means that these two words, which are assigned factors themselves, have a higher frequency of occurrence than any other assigned factors, as seen in Fig. 7.2. Hence, more attention should be focused on these two factors due to their significant occurrence among all 20 interview transcripts that were initially

Fig. 7.1 Results in word clouds

Category	Code / Factor	Description	Count	% Codes	Cases	% Cases
Subcontracting Practices\Characteristics\Flexibility	Resource deployment		64	10.9%	20	100.0%
Outcomes	Improved Work Productivity		58	9.9%	20	100.0%
Subcontracting Practices\Characteristics\Flexibility	Work Arrangement		52	8.9%	18	90.0%
Outcomes	Decreased Work Productivity		46	7.8%	18	90.0%
Subcontracting Practices\Perceived Benefits	Efficient		43	7.3%	18	90.0%
Subcontracting Practices\Characteristics\Flexibility	Enables Resource Control		42	7.2%	16	80.0%
Outcomes	Increased Resource Wastage		41	7.0%	14	70.0%
Actors\External Forces	Change in Govt. Policies		28	4.8%	17	85.0%
Subcontracting Practices\Characteristics\Coordination Levels	Ample coordination amongst parties		28	4.8%	15	75.0%
Actors\External Forces	Industry Irregularities		25	4.3%	13	65.0%
Subcontracting Practices\Characteristics\Coordination Levels	Lack of coordination amongst parties		22	3.8%	13	65.0%
Actors\Subcontractors	Specialised work		21	3.6%	13	65.0%
Subcontracting Practices\Perceived Benefits	Achieves Sustained Competitive Advantage		19	3.2%	13	65.0%
Subcontracting Practices\Characteristics\Flexibility	Resource sharing		13	2.2%	9	45.0%
Subcontracting Practices\Characteristics\Multiple Tiers	Subcontracting/Outsourcing by MC to SC		12	2.0%	10	50.0%
Subcontracting Practices\Characteristics\Multiple Tiers	Subcontracting/Outsourcing from SC to SC		11	1.9%	11	55.0%
Actors\External Forces	Economic Downturn		11	1.9%	8	40.0%
Actors\Subcontractors	Small Employment size		10	1.7%	7	35.0%
Outcomes	Reduced Resource Wastage		10	1.7%	5	25.0%
Actors\Subcontractors	Low construction turnover		9	1.5%	6	30.0%
Subcontracting Practices\Perceived Benefits	Accrue Firm Flexibility		7	1.2%	7	35.0%
Actors\Main Contractors	Undertake General work		5	0.9%	5	25.0%
Actors\Subcontractors	Small scale work		4	0.7%	3	15.0%
Actors\Main Contractors	High construction turnover		2	0.3%	2	10.0%
Actors\Main Contractors	Large scale work		2	0.3%	1	5.0%
Actors\Main Contractors	Large Employment Size		1	0.2%	1	5.0%
Actors\External Forces	Globalisation					

Fig. 7.2 Results in tabular form

coded prior to the analysis by the software. This was the starting point of the analysis. Other notable word clouds include Work Arrangement, Decreased Work Productivity, Increased Resource Wastage, Efficient, and Enables Resource Control, which are circled in blue.

A further examination of the screenshot shown in Fig. 7.2 explains the significance of the word clouds Resource Deployment and Improved Work Productivity. The frequency count, which is a measure of the number of times that an event occurs for both Resource Deployment and Improved Work Productivity, stands at 64 and 58, respectively. In addition, these two assigned factors have appeared in all 20 transcripts. As shown in Fig. 7.2, both factors are found in all 20 cases.

A visual perspective, in the form of a horizontal bar chart, is further drawn up to provide better visual representation of the distribution of code frequency. This is shown in Fig. 7.3.

From Fig. 7.3, no doubt that the highest code frequency is Resource Deployment, followed by Improved Work Productivity. As mentioned earlier, other significant factors are Work Arrangement, Decreased Work Productivity, Increased Resource Wastage, Efficient, and Enables Resource Control.

However, one glaring discrepancy would be the code named Globalisation, where there is no count or frequency associated with it in all transcripts as highlighted in yellow. This is also observed in Fig. 7.2 where there is no value attributed to the frequency factor Globalisation as highlighted in yellow. A possible deduction could be that all 20 interviewees do not view the effect of globalisation as an immediate threat to their livelihood and to the construction industry. It is also possible that the interviewees have little or no knowledge on the topic of globalisation.

Nonetheless, through these results, it can be inferred that factors Resource Deployment and Improved Work Productivity are key contributing factors to explain the prevalent use of subcontracting practices, as their factors surfaced throughout all 20 interview transcripts. Based on the coding analytical framework as depicted in Chap. 6, such an occurrence can be attributed to the flexibility given to construction contractors that is desired when they subcontract their work out to other contractors down the MLSS hierarchy, which perhaps affects productivity in a positive manner. Hence, this inference warrants further investigation to determine any significant relationship between these two factors. Other notable factors will also be scrutinised in the same way as the factors Resource Deployment and Improved Work Productivity to determine how these factors influence the reasons for subcontracting and its subsequent effect on construction productivity.

Figure 7.4 shows the frequency distribution of keywords or factors when QDA Miner generated these results. From this graphical representation, it depicts multiple factors that were factored into the analysis based on the input of the proposed coding analytical framework. Almost all factors, except Globalisation, occurred more than once.

Due to such overwhelming results, there is a need to further select factors that have a higher frequency of occurrence to enable a more focused analysis of factors that have higher influence than others on the reasons for contractors' preferences for

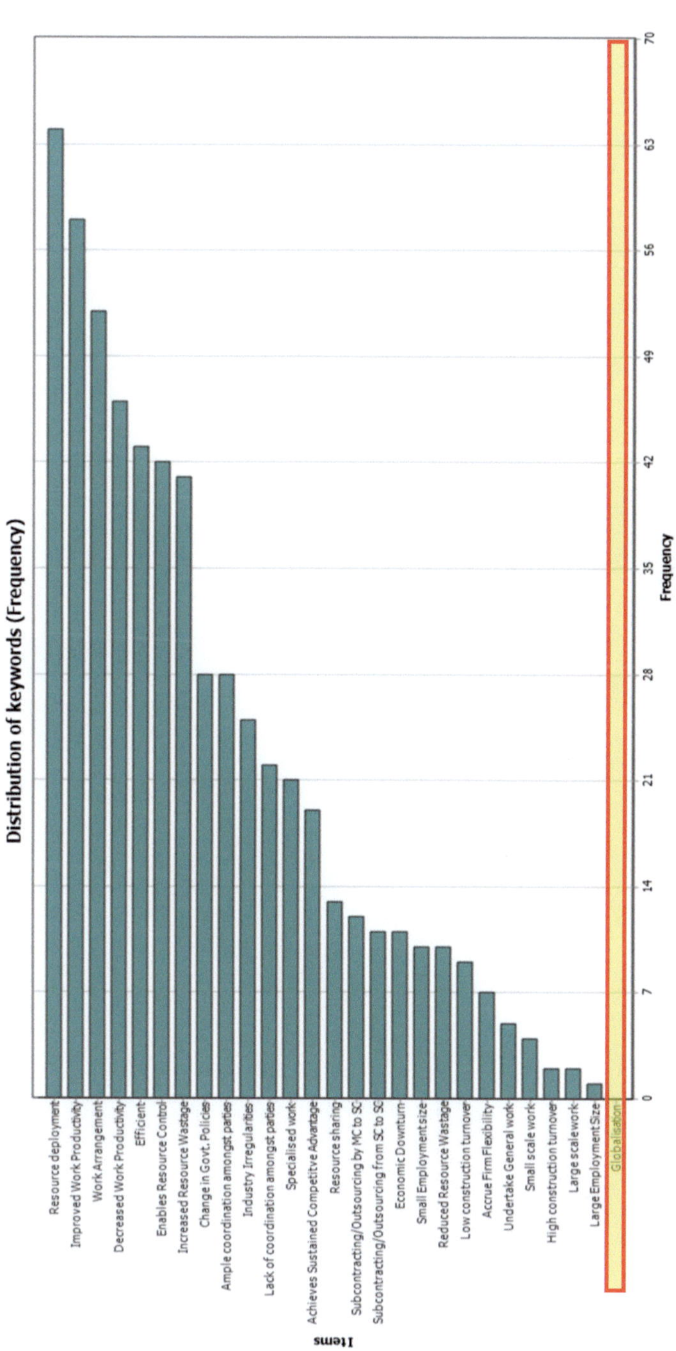

Fig. 7.3 Results in horizontal bar chart

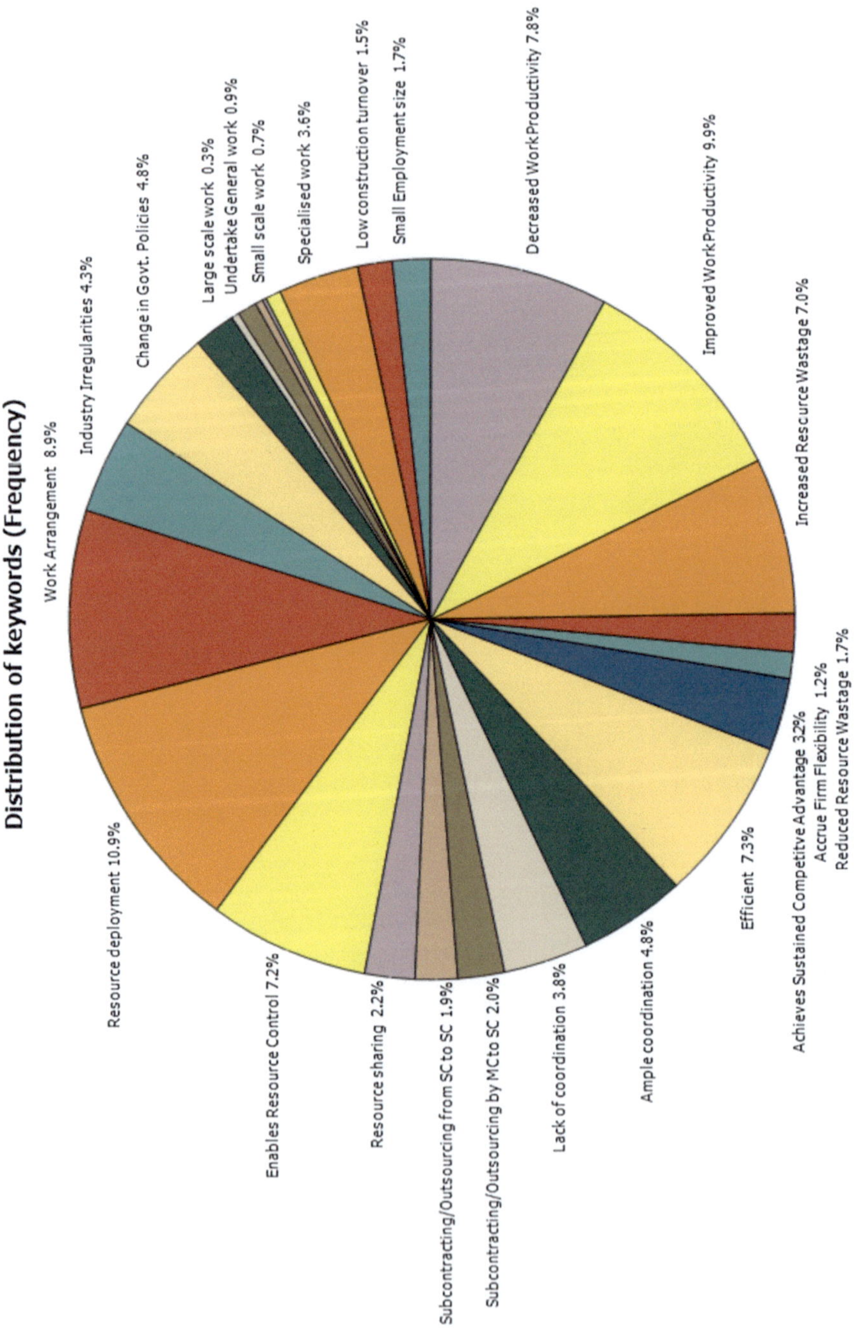

Fig. 7.4 Results in pie chart

subcontracting and its overall effect on productivity. The more prevalently occurring factors are presented as a percentage occurrence out of the total number of factors present, as follows:

- Resource Deployment—10.9%,
- Improved Work Productivity—9.9%,
- Work Arrangement—8.9%,
- Decreased Work Productivity—7.8%,
- Efficient—7.3%,
- Enables Resource Control—7.2%,
- Change in Government Policies—4.8%, and
- Ample Coordination—4.8%.

7.2 Cluster Analysis

The next set of results are obtained based on the adoption of the hierarchical cluster analysis. This analysis is an agglomerative methodology that identifies clusters of observations in a data set (Aldenderfer & Blashfield, 1984). Anderberg (1973) described the cluster analysis at its most elementary level, where it "sorts data into groups, such that the degree of natural association is high among members of the same group and low between members of different groups".

Basically, a cluster analysis is a statistical method for grouping like things together. Before performing a cluster analysis, there is a need to display qualitative data using similarity matrices. A similarity matrix is a table of scores that express numerical distances—or likeness—between data points. These matrices can be weighted (using a range of values to indicate the degree or similarity strength) or binary (using ones and zeroes to indicate simply whether two data points are similar or not). These matrices can be composed of data from various sources, including direct word counts or code frequencies, depending on which item of analysis best addresses the analysis objectives. Hence, by adopting the similarity matrices as input, cluster analysis allows a researcher to see patterns in large data sets. Thus, a refined context can be created by careful selection of the words or factors to be included in a similarity matrix.

As such, choosing an appropriate software package for a particular analysis task is contingent upon the researchers' analysis objectives as well as the size and complexity of both the data set and the analysis team (MacQueen, 2005). Hence, the employment of the QDA software technology was vital, as it is able to fulfil the requirements of conducting a hierarchical cluster analysis. Based on the 20 interview transcripts, a hierarchical cluster analysis is adopted, with the analysis results presented in the form of a dendrogram, as shown in Fig. 7.5.

As observed in Fig. 7.5, the result of the hierarchical cluster analysis is a dendrogram. A dendrogram is defined as a branching diagram that represents the relationships of similarity among a group of entities (Wheaton College, 2016). It

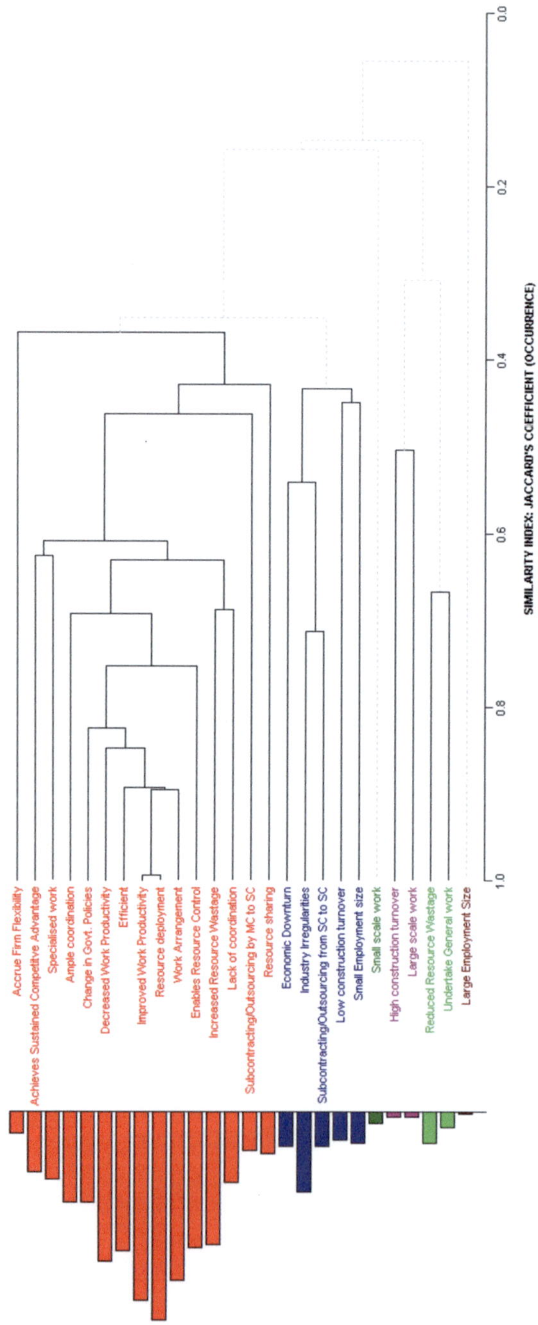

Fig. 7.5 Dendrogram results

shows the links between factors, and its structure offers clues as to which factors form coherent clusters. Therefore, the dendrogram provides a starting point from which to develop a more comprehensive interpretation of the data.

As shown in Fig. 7.6, the main features of a dendrogram are the following:

1. Each branch is called a *clade/node* (denoted with purple arrows);
2. The terminal end of each branch is called a *leaf* (denoted with brown arrows); and
3. Branches can have just one leaf, called simplicifolious or 'single-leafed'. This means that its placement is an indication that the distribution of factors in that chunk is substantially different from the distribution in the remaining clusters.

7.2.1 Key Features of the Dendrogram

See Fig. 7.6.

7.2.2 How to Interpret the Dendrogram

In Fig. 7.7, clusters, which are defined as individual segments of factors that are most similar to each other, are grouped together. For instance, the two most similar clusters are combined and continue to combine until all objects are in the same cluster. As such, hierarchical clustering produces a tree (called a dendrogram) that shows the hierarchy of the clusters. This allows for exploratory analysis to see how the microarrays group together based on similarity of features (Partek, 2012). Within a cluster, there are also factors that occur at a lower rate while others occur at higher rates. The factors that are grouped at higher rates suggest co-occurrence as well as high frequency within the text. As such, it is possible to isolate clusters of factors with high similarity, thereby enabling an in-depth analysis of these factors.

There are four different clusters and two un-clustered factors in Fig. 7.7. Details are shown below:

- The portion boxed in red is known as Cluster 1;
- The portion boxed in blue is known as Cluster 2;
- The portion boxed in purple is known as Cluster 3;
- The portion boxed in green is known as Cluster 4; and
- The portions highlighted in yellow are un-clustered factors.

In Fig. 7.7, the x-axis shows the similarity index using Jaccard's coefficient, where it determines the association between two factors (Niwattanakul, Singthongchai, Naenudorn, & Wanapu, 2013). The higher the index, the higher the similarity between the factors. In addition, the Jaccard index is a term often used for

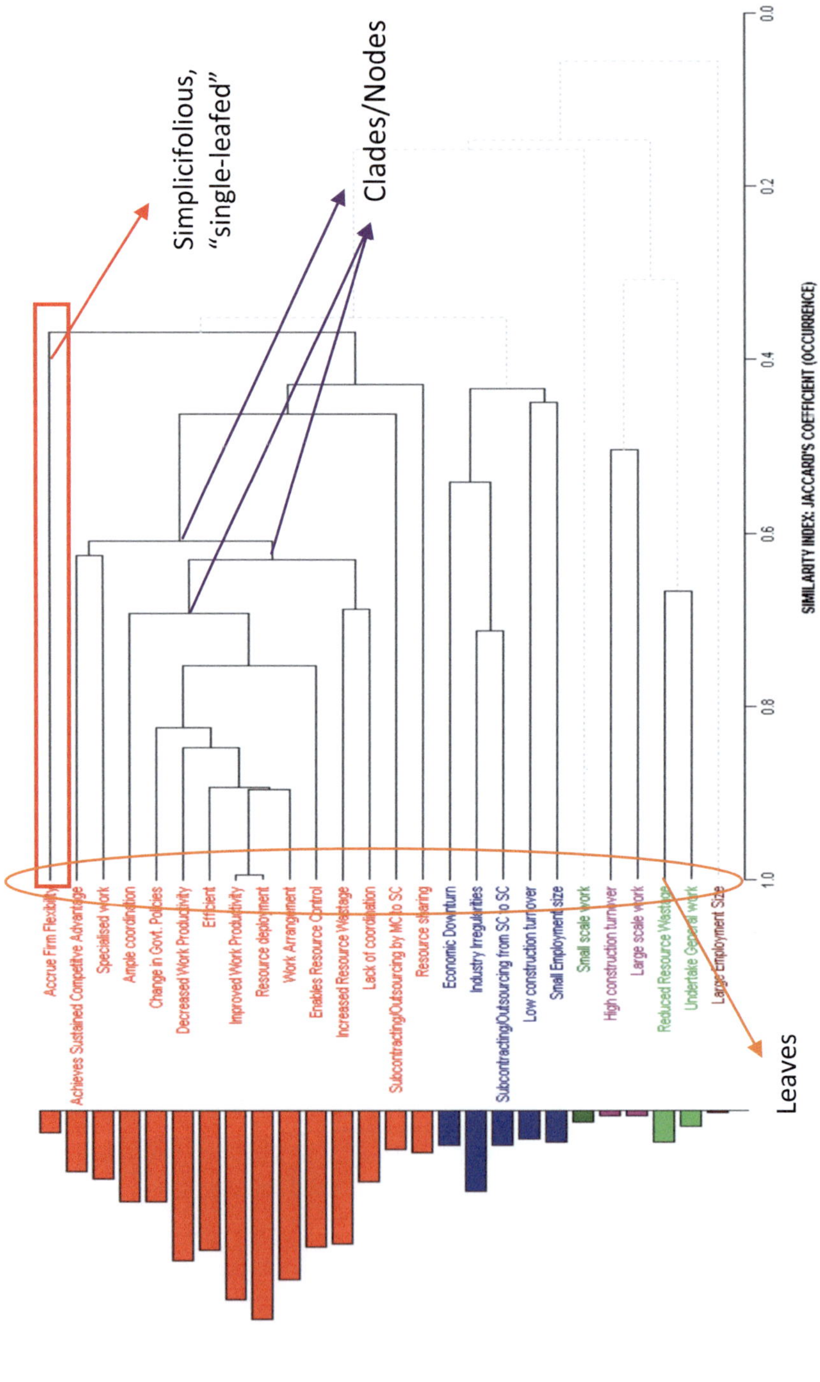

Fig. 7.6 Key features of the dendrogram

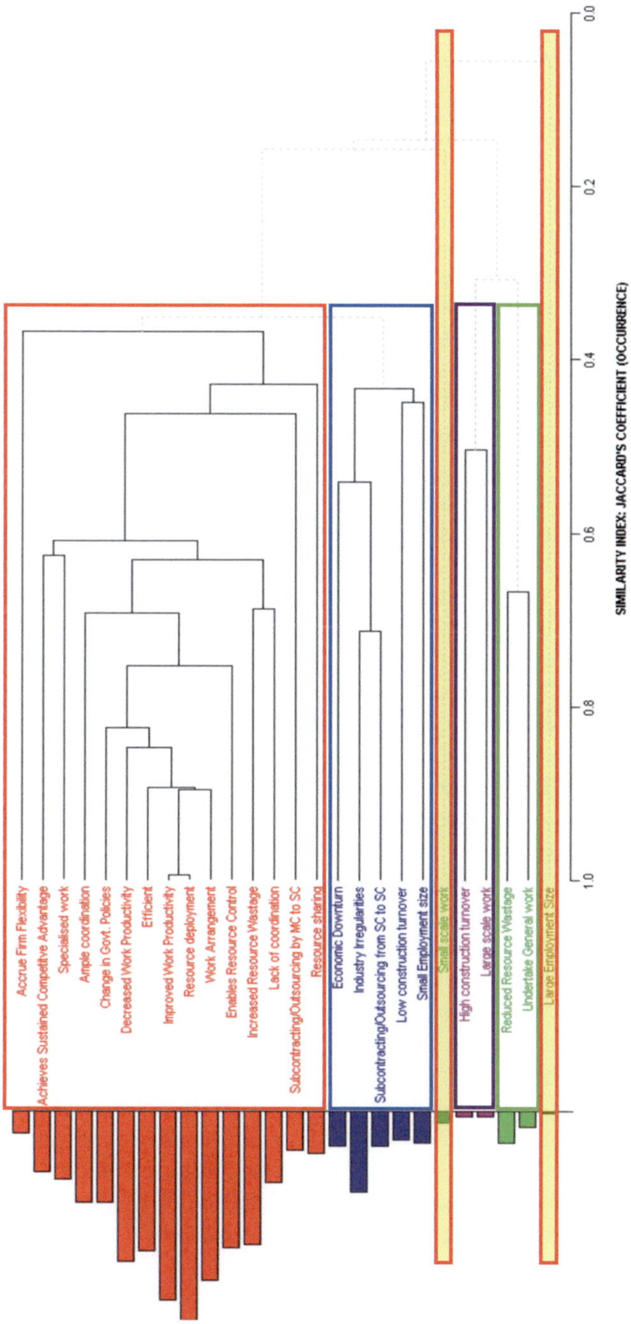

Fig. 7.7 Grouping of dendrogram results

comparing similarity, dissimilarity, and distance of the data set of the code. The Jaccard similarity coefficient between two data sets/factors is derived by the division between the number of features that are common to all divided by the number of properties. The formula is shown below:

$$J(A, B) = \frac{|A \cap B|}{|A \cup B|}$$

As observed, factors that are in Cluster 1 (red) bear more similarity than other clusters, as the cluster is the largest in terms of the number of factors and has the most factors with a higher similarity index of close to 1. To further verify this observation, the QDA Miner software has an analysis function called 'Coding Co-occurrences—Similarity Table'. This function produces a table that computes the similarity of one assigned code with the other, and this ultimately allows the deduction of the relationships of similarity among the group of factors in this study, which enables the construction of the dendrogram. The similarity table is shown in Fig. 7.8.

Figure 7.8 shows the similarity index of one code with the other. In each coloured box is a computed similarity index, the lighter the colour of the box, the higher the similarity index will be between the two factors and vice versa. As shown, the similarity index between factors Resource Deployment and Improved Work Productivity is 1, with the brightest colour of yellow in the grid, which is boxed in red. This means that there is extremely strong similarity between these two factors and hence a strong relationship between these two factors, where the occurrence of one will affect the other. Conversely, the grids of darker colour imply that the two factors have lower similarity index, especially those in black, where the value of zero implies no similarity between the two assigned factors.

To enable the ease of deciphering the relationship among various factors via the similarity index, the software has a function called *agglomeration*, where it has tabulated the similarity index of all factors in the analysis. This produces an agglomeration table where it ranked the factors from the highest similarity index to the lowest. The screenshot of the agglomeration table is as shown in Fig. 7.9.

From Fig. 7.9, it was tabulated that the assigned factors Improved Work Productivity and Resource Deployment have a similarity index of 1, which is also depicted in Fig. 7.8. Moreover, the advantage of an agglomeration is that it lists factors that have the highest similarity index, also known as relationships of similarity between factors, for ease of further analysis. Here, this study will analyse the top six nodes that are boxed in red in Fig. 7.9 to enable a more focused approach in achieving the research outcome. This is also, in part, because of their high similarity index of 0.75–1.00, which is more feasible in determining the reasons contractors predominantly adopt subcontracting practices and the effect of such practices on construction productivity.

Hence, more focus will be on Cluster 1 regarding the factors within the top six nodes, as identified in Fig. 7.9. This is highlighted in yellow in Fig. 7.10, which consists of all factors grouped within Cluster 1.

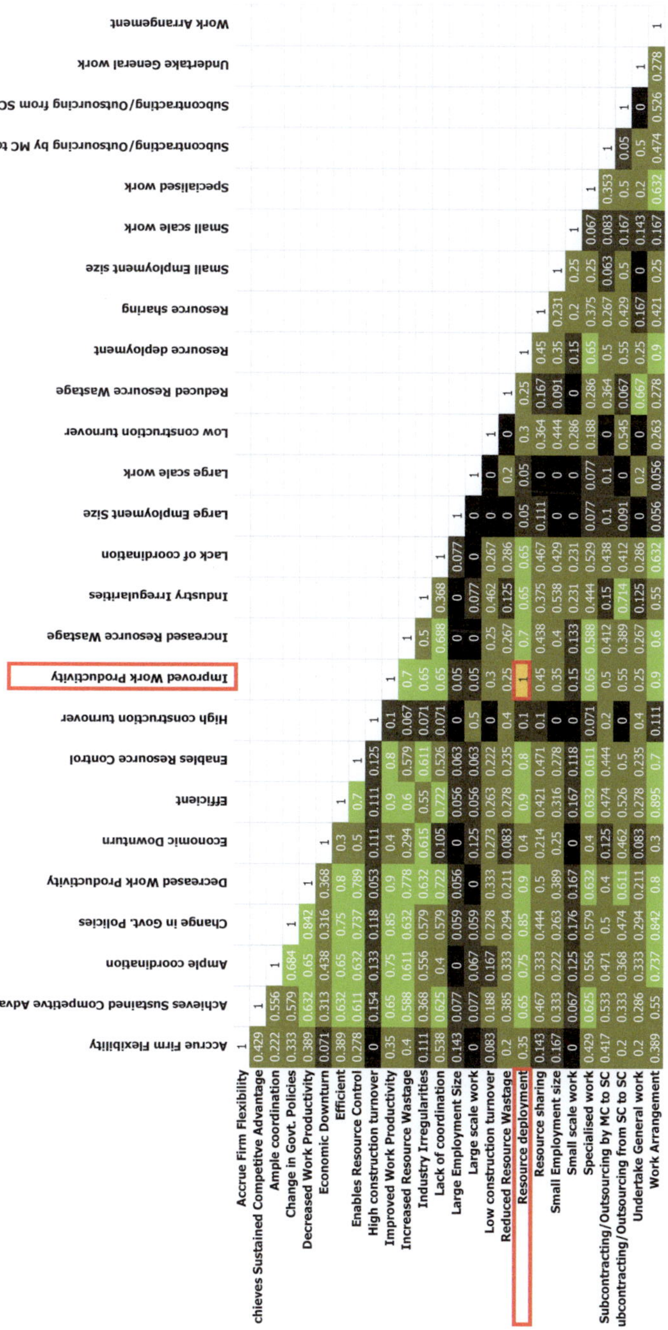

Fig. 7.8 Similarity index diagram

Node	Group 1	Group 2	Similarity
1	Improved Work Productivity	Resource deployment	1.000
2	Node 1	Work Arrangement	0.900
3	Efficient	Node 2	0.898
4	Decreased Work Productivity	Node 3	0.850
5	Change in Govt. Policies	Node 4	0.827
6	Node 5	Enables Resource Control	0.754
7	Industry Irregularities	ıbcontracting/Outsourcing from SC to	0.714
8	Ample coordination	Node 6	0.693
9	Increased Resource Wastage	Lack of coordination	0.688
10	Reduced Resource Wastage	Undertake General work	0.667
11	Node 8	Node 9	0.630
12	chieves Sustained Competitve Advanta	Specialised work	0.625
13	Node 12	Node 11	0.607
14	Economic Downturn	Node 7	0.538
15	High construction turnover	Large scale work	0.500
16	Node 13	;ubcontracting/Outsourcing by MC to S	0.458
17	Low construction turnover	Small Employment size	0.444
18	Node 14	Node 17	0.428
19	Node 16	Resource sharing	0.423
20	Accrue Firm Flexibility	Node 19	0.361
21	Node 20	Node 18	0.344
22	Node 15	Node 10	0.300
23	Node 21	Small scale work	0.147
24	Node 23	Node 22	0.136
25	Node 24	Large Employment Size	0.043

Fig. 7.9 Screenshot of agglomeration table

As noticed on the far-left of Fig. 7.10, there is a horizontal bar chart in red. In QDA Miner 4.0, the software has a display function of enabling the clustering of factor co-occurrences, where bars are displayed beside each dendrogram item to represent their relative frequencies. This provides a clearer association, in terms of relationships, between each code in the dendrogram, their frequency of occurrence within the 20 transcripts, and how they relate to another similar code that it is being grouped within the dendrogram.

7.2.3 Coding Co-occurrences

Another term that is mentioned frequently in cluster analysis is the term "code co-occurrence". It is defined as the application of two or more factors to a discrete segment of text from a unique respondent (Guest & McLellan, 2003). For instance, a text paragraph from one respondent's interview transcript about the reason the firm adopts subcontracting as one of its strategies may contain references to Resource Deployment, Enables Resource Control, Work Arrangement, and

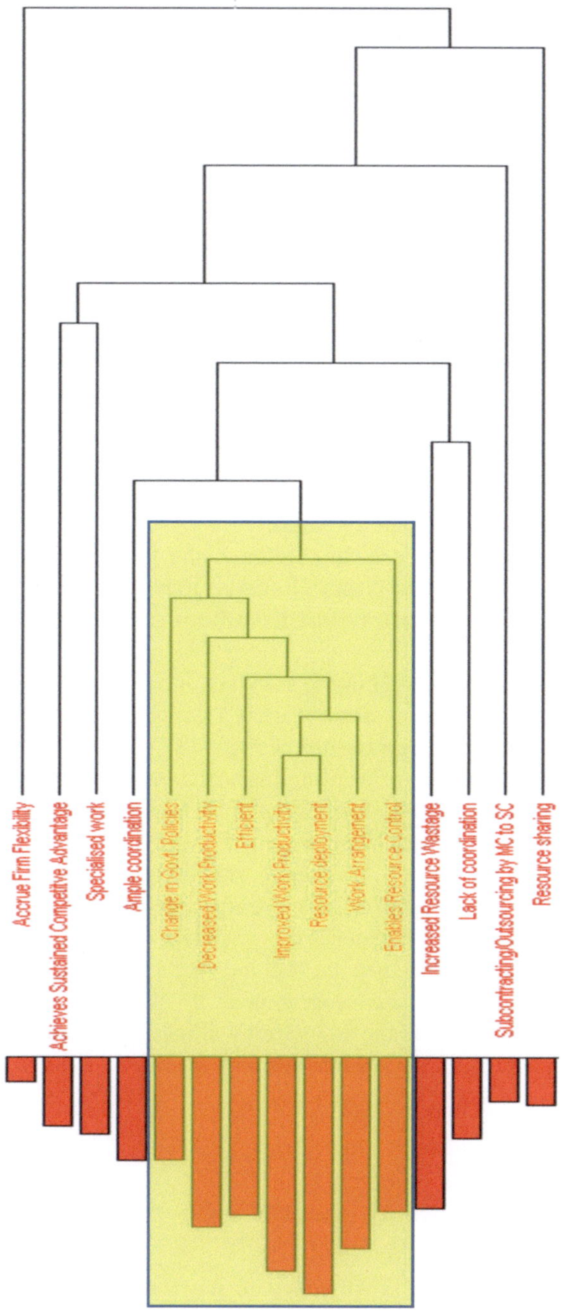

Fig. 7.10 Cluster 1 of dendrogram results

Efficient. Each of these references is a separate idea that requires a separate code. The paragraph would be coded with all three factors, which are then said to co-occur in this text segment. The reports of code co-occurrence often provide helpful information in understanding how thematic domains, concepts, or ideas are distributed within a data set, beyond simple frequencies. Patterns in data sets can be identified by examining co-occurrences, such as correlation between code items, themes, respondents, or events (LeCompte & Schensul, 1999). For coding co-occurrences, the sequence of how the factors occur does not matter. This is provided in the software in the form of a co-occurrences table, as shown in Fig. 7.11.

Figure 7.11 operates similarity to Fig. 7.8; the lighter the colour of the grid, the higher the co-occurrence frequency of the two factors that are computed and vice versa. In this case, code Resource Deployment and code Improved Work Productivity have the highest co-occurrence frequency of 20, with its grid highlighted in yellow. This means that the factor Resource Deployment occurs within the same paragraph text with factor Improved Work Productivity 20 times throughout the 20 interview transcripts. Such an occurrence can be deduced as a close proximity or relationship among these two factors, as they both keep occurring together for a significant 20 times.

The grids that are highlighted in blue refer to the frequency of occurrence of each factor throughout the 20 transcripts. However, to better focus on this study, we will only focus on the seven factors that have been earlier identified. These are boxed in red in Fig. 7.11.

Nonetheless, like any such data-mining software, it has weaknesses. An anomaly is detected, where boxed in red, it shows factors Decreased Work Productivity and Efficient as having a rather high co-occurrence value of 16. Such an occurrence can be flagged as a typical weakness of data-mining software where it is programmed to embark on the sequencing of these factors based on how the text is coded. It could very well be that the interviewees mentioned that, in order to minimise a decline in work productivity, the firms must work more efficiently. This leads to the factors Decreased Work Productivity and Efficient being coded one after another, which caused the anomaly. Therefore, it exposes the weakness of such software in that it is unable to fully comprehend the meaning behind these texts but can only operate based on coded sections that are programmed by the user. Hence, it is still imperative that the analyst guide the software by allocating appropriate sections of text with the relevant codes to enable optimal results to be derived from the software.

7.2.4 Proximity Analysis by Jaccard's Coefficient Index

In addition, proximity analysis is employed in the form of Jaccard's coefficient index to measure proximity of the numerous factors that are input into the content-analysis software. QDA Miner 4.0 was able to gauge the levels of

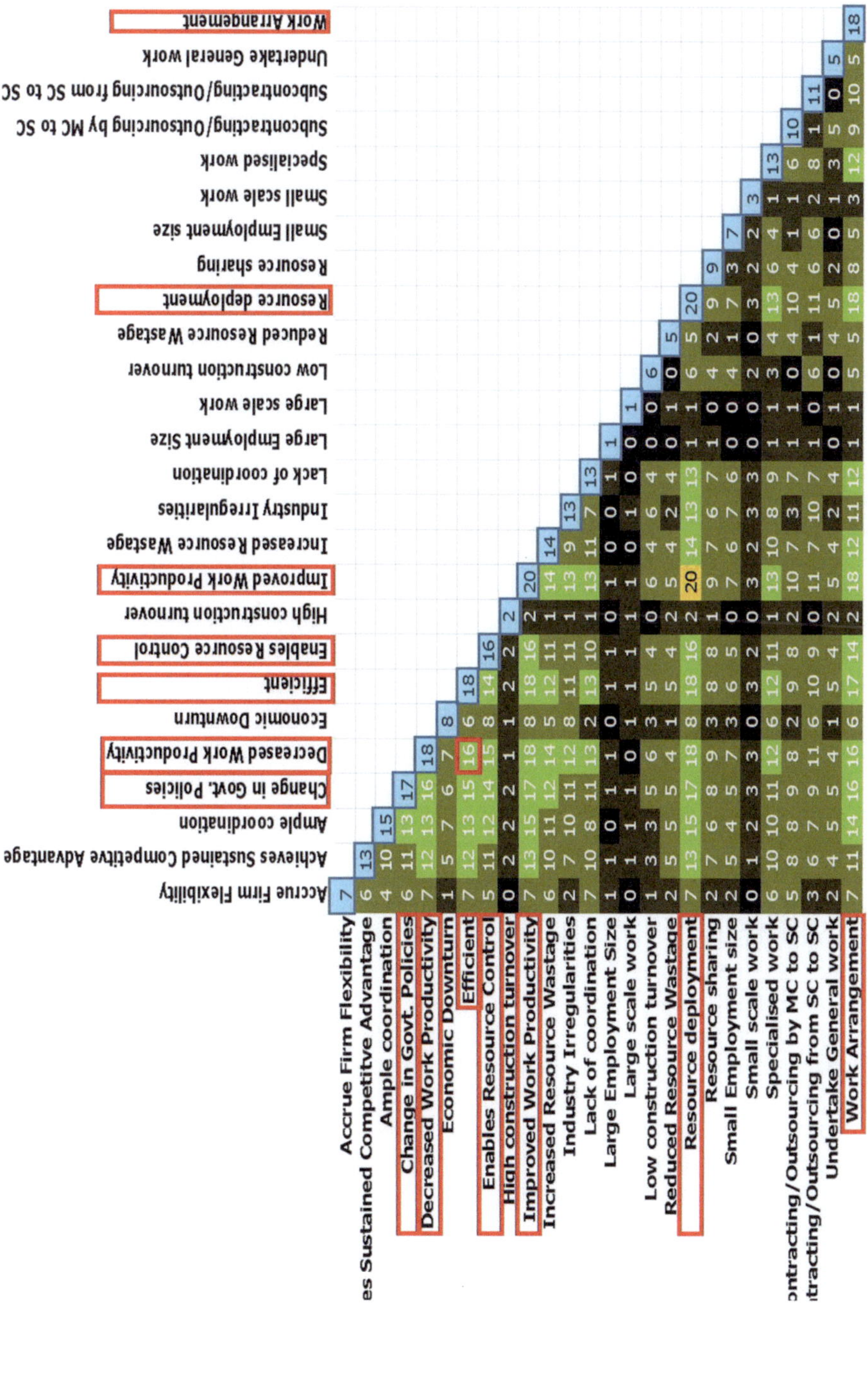

Fig. 7.11 Co-occurrences table

proximities between the assigned factors. Jaccard's coefficient was chosen because Lugwig and Reynolds (1988) and Hubalek (1982) experimented with the coefficient and agreed with its admissibility to compare the proximity and diversity of sample sets. The maximum possible value of a Jaccard's coefficient between two factors is 1. This means that the closer the value between the two factors is to 1, the more proximate these two factors will be. With that, a proximity analysis is conducted on the seven chosen factors. A screenshot of the results from the proximity analysis, pertaining to one of the chosen factors, is shown in Fig. 7.12.

As observed in Fig. 7.12, the chosen code is Resource Deployment, hereby referred to as the selected code, which is categorised under the column 'Target'. The subsequent column 'Keyword' contains the code Improved Work Productivity, which is hereby referred to as the given code. It is noted that both factors have the highest possible proximity value of 1.0, which means that these two factors are very proximate or share an extremely close relationship.

In addition, Fig. 7.12 allows a detailed examination of the numerical values behind the calculation of these plots. When the distance measure is based on co-occurrences, the table provides detailed information, such as the number of times a given code co-occurs with the selected one (CO-OCCURS) and the number of times that the given code appears in the absence of this selected code (DO NOT). The table also includes the number of times the selected code appears in the absence of the given code (IS ABSENT).

Moreover, the selected code Resource Deployment, with its corresponding given code Improved Work Productivity has 20 instances of co-occurrences, the highest among the other code combinations. This is where the given code Improved Work Productivity co-occurs with the selected code Resource Deployment. Neither has Improved Work Productivity appeared in the absence of Resource Deployment nor has Resource Deployment appears in the absence of code Improved Work Productivity. This reflects the very strong proximity and therefore strong relationship between these two factors as they always co-occur.

However, to enable a more refined and focused analysis from the seven chosen factors, only the top result from each factor will be displayed in the form of a tabulated results table. This is as shown in Table 7.2.

Table 7.2 shows a tabulation of the key results from the proximity analysis, where the leftmost column of Table 7.2 consists of the seven chosen factors to be analysed. The subsequent column shows the corresponding factors (given factors) that have the closest proximity to the chosen factors (selected factors) based on the Jaccard's coefficient index. These factors have a very high Jaccard's index of between 0.800–1.000 and the two most notable given factors are Resource Deployment and Improved Work Productivity. This shows that these two factors are the most significant among all other assigned factors, which points to the possibility that the key reason for the prevalence of subcontracting practices in the local industry is due to the flexibility given to firms via Resource Deployment, where upon doing so, such firms will achieve an outcome of Improved Work Productivity. Nonetheless, other forms of analysis will be adopted to further validate such a preliminary conclusion.

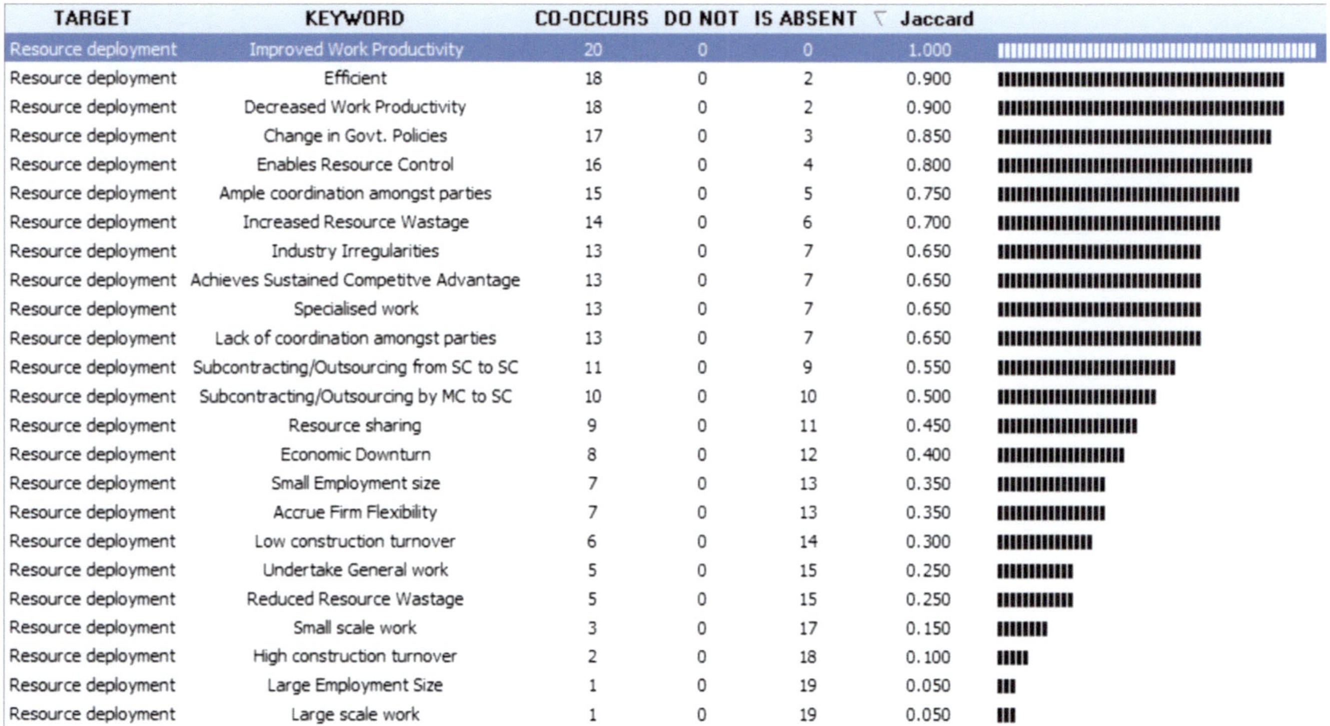

TARGET	KEYWORD	CO-OCCURS	DO NOT	IS ABSENT	Jaccard
Resource deployment	Improved Work Productivity	20	0	0	1.000
Resource deployment	Efficient	18	0	2	0.900
Resource deployment	Decreased Work Productivity	18	0	2	0.900
Resource deployment	Change in Govt. Policies	17	0	3	0.850
Resource deployment	Enables Resource Control	16	0	4	0.800
Resource deployment	Ample coordination amongst parties	15	0	5	0.750
Resource deployment	Increased Resource Wastage	14	0	6	0.700
Resource deployment	Industry Irregularities	13	0	7	0.650
Resource deployment	Achieves Sustained Competitve Advantage	13	0	7	0.650
Resource deployment	Specialised work	13	0	7	0.650
Resource deployment	Lack of coordination amongst parties	13	0	7	0.650
Resource deployment	Subcontracting/Outsourcing from SC to SC	11	0	9	0.550
Resource deployment	Subcontracting/Outsourcing by MC to SC	10	0	10	0.500
Resource deployment	Resource sharing	9	0	11	0.450
Resource deployment	Economic Downturn	8	0	12	0.400
Resource deployment	Small Employment size	7	0	13	0.350
Resource deployment	Accrue Firm Flexibility	7	0	13	0.350
Resource deployment	Low construction turnover	6	0	14	0.300
Resource deployment	Undertake General work	5	0	15	0.250
Resource deployment	Reduced Resource Wastage	5	0	15	0.250
Resource deployment	Small scale work	3	0	17	0.150
Resource deployment	High construction turnover	2	0	18	0.100
Resource deployment	Large Employment Size	1	0	19	0.050
Resource deployment	Large scale work	1	0	19	0.050

Fig. 7.12 Results of proximity analysis

Table 7.2 Tabulated results table on proximity analysis

Target (Selected factor)	Keyword (Given factor)	Co-occurs	Do not	Is absent	Jaccard's coefficient index
Improved work productivity	Resource deployment	20	0	0	1.000
Resource deployment	Improved work productivity	20	0	0	1.000
Work arrangement	Improved work productivity	18	2	0	0.900
Efficient	Resource deployment	18	2	0	0.900
Decreased work productivity	Resource deployment	18	2	0	0.900
Change in government policies	Improved work productivity	17	3	0	0.850
Enables resource control	Improved work productivity	16	4	0	0.800

7.3 Frequency Matrix Analysis

A frequency matrix is a table of scores that expresses the number of times (frequency) a code follows another code in the data points. These matrices are weighted (using values to indicate the number of times it co-occurs). These matrices can be composed of data from various sources, including direct word counts or code frequencies, depending on which item of analysis best addresses the analysis objectives. Using these frequency matrices as input, the cluster analysis allows the analyst to see patterns in large data sets. Moreover, a refined context can be created by careful selection of the words or factors to be included in a frequency matrix. If the words or factors are limited to those that appear within a particular domain or structural-level code segment, a bounding context can be used in interpreting the resulting clusters. It is important to note that this analysis is used to identify factors that not only co-occur but also do so in a specific order and under specific conditions. The adoption of the frequency matrix analysis has produced the result as shown in Fig. 7.13.

As depicted in Fig. 7.13, factors that are on the left portion of the figure are grouped as A, whereas the factors that are on the top portion of the figure are grouped as B. Figure 7.13 serves to show the relationships between any two factors, in terms of the sequence of occurrence within all 20 transcripts. As observed, the highest frequency of one code that follows the other is where Resource Deployment (B) follows Work Arrangement (A), as the grey square indicates it occurs 28 times. This seems to suggest that contractor firms typically discuss and agree on certain work arrangements first, prior to deploying their firm's resources where, based on

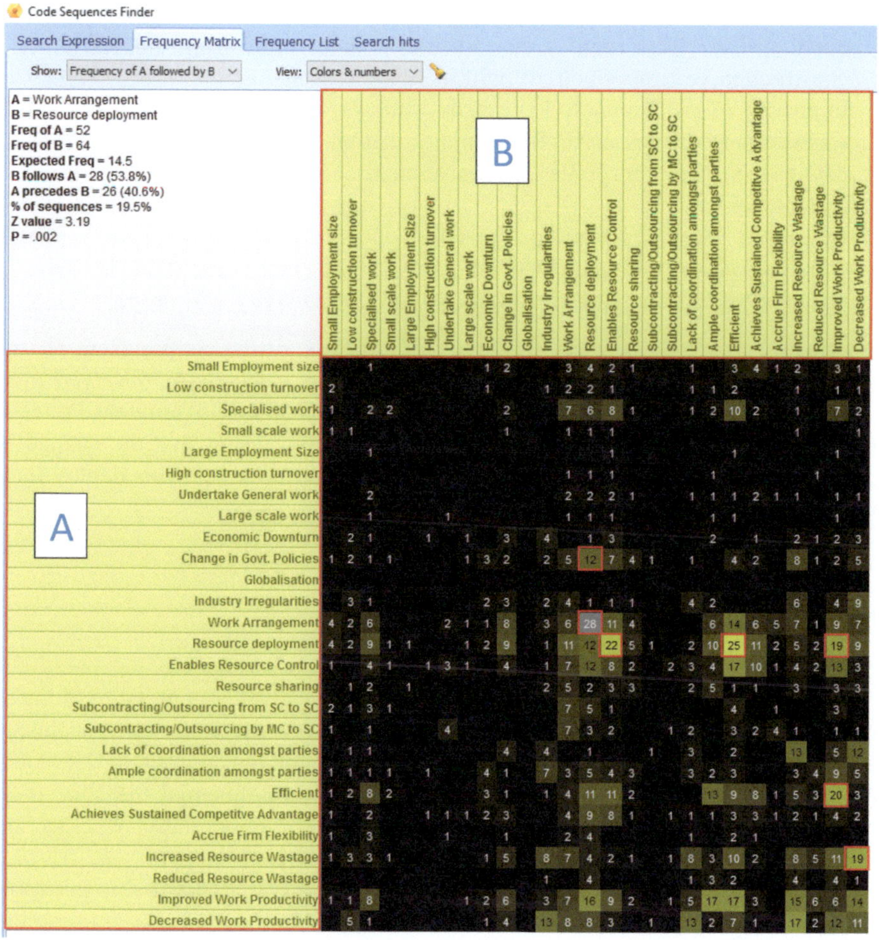

Fig. 7.13 Frequency matrix analysis results

the agreed work arrangements, the firms would then know what resources to deploy and how much to deploy.

Other prominent frequency sequences among the factors are as follows:

1. Efficient typically follows Resource Deployment, which stands at a frequency of 25. This might imply that the types of resources that are deployed at the work site by the contractors will affect the efficiency of the job.
2. Enables Resource Control generally follows Resource Deployment, which stands at a frequency of 22. This implies that the way in which contractor firms deploy their resources will help them accrue savings if Resource Deployment is carried out properly.

3. Improved Work Productivity mostly follows Efficient, which stands at a frequency of 20. This implies that when contractor firms work efficiently, their construction work productivity will increase as well.
4. Improved Work Productivity mostly follows Resource Deployment, which stands at a frequency of 19. This implies that the way in which contractor firms deploy their firm's resources will cause their construction work productivity to increase as well.
5. Decreased Work Productivity typically follows Increased Resource Wastage, which stands at a frequency of 19. This implies that when there is a rise in resource wastage on site, construction work productivity will drop.

However, there is a notable low frequency for one of the six prevalent factors that this analysis focused on—Change in Government Policies. As observed from Fig. 7.13, Resource Deployment mostly follows Change in Government Policies, which only stands at a frequency of 12. This probably means that the change in government policies does not have much influence on how the firms deploy their resources, which could explain the low frequency.

7.4 Pearson's Correlation Analysis

Pearson's product moment correlation coefficient (r) is given as a measure of linear association between two variables (StatsDirect, 2016), where it shows how closely related one code is with the other code. This will enable the analysis to determine how much influence a code has on other factors, how influential the code is in explaining the prevalence of subcontracting, and most importantly, whether the code has an effect on construction productivity.

To aid in the correlation analysis, the QDA Miner 4.0 software is able to export the data results to other statistical software package like the SPSS, by framing the data in Boolean terms. This data export process is shown in the boxed section of Fig. 7.14. As shown in Fig. 7.14, the export of the results in Boolean terms will enable a thorough analysis to be conducted with regards to correlation between the various factors using SPSS.

7.4.1 Pearson's Correlation Analysis via SPSS

The exported results from QDA Miner 4.0 were input into SPSS, with SPSS supporting its file format. As reiterated earlier, the correlation and regression analysis via SPSS will cover seven assigned factors that have higher frequency occurrence throughout the 20 interview transcripts based on other forms of analysis. The SPSS results with respect to the seven factors are shown in Fig. 7.15.

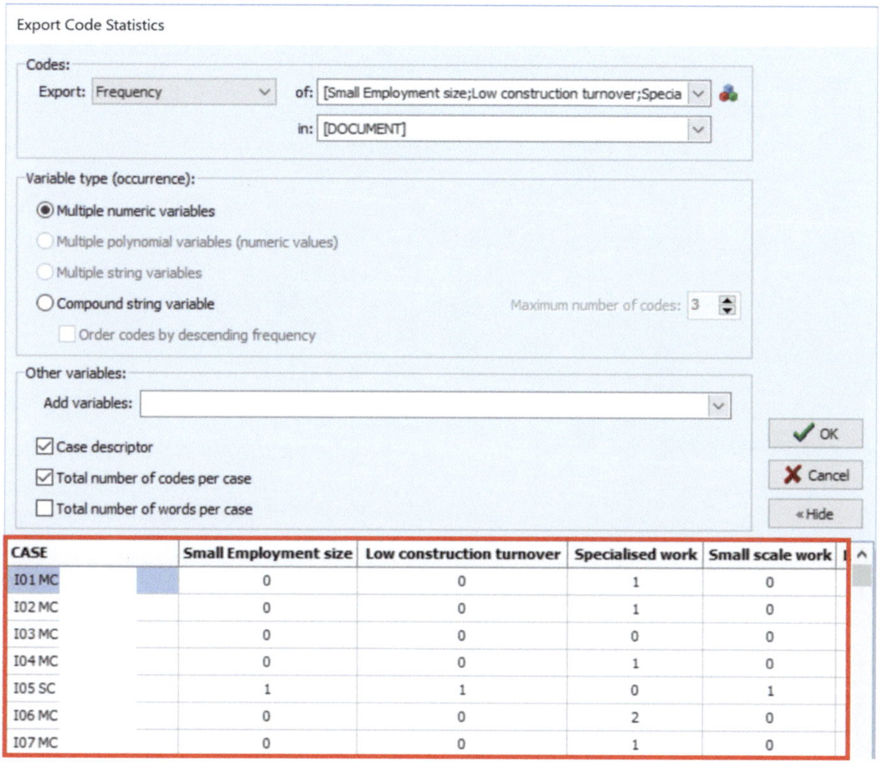

Fig. 7.14 Screenshot of data export process from QDA Miner 4.0

As observed in Fig. 7.15, the seven chosen factors with their correlation values with respect to the other factors, which are statistically significant (two-tailed test) at the 0.05 level have been highlighted in green. Therefore, it can also be deduced that such a correlation between these factors does not occur randomly and, as such, brings reliability to the results. Moreover, as the same *p*-value for correlation analysis is also derived from the regression analysis, a statistical correlation analysis would suffice. Such an analysis indicates the confidence of the correlation with each individual factor and the dependent factor. Most importantly, the *p*-value also ascertains the validity of the relationship.

In addition, the assigned factor Globalisation in Fig. 7.15, which is boxed in red, depicts no correlation with any other assigned factors. This is because none of the interviewees mentioned any word, phrase, or reference to globalisation during the interviews. This could imply that the effects of globalisation are not yet of obvious concern to contractors or that they may not even understand the term *globalisation* due to lack of awareness or lower education levels.

Correlations

		Small Employment size	Low construction turnover	Specialised work	Small scale work	Large Employment Size	High construction turnover	Undertake General work	Large scale work	Economic Downturn	Change in Govt Policies	Globalisation	Industry Irregularities	Work Arrangement	Resource deployment	Enables Resource Control	Resource sharing	Subcontracting /Outsourcing from SC to SC	Subcontracting /Outsourcing by MC to SC	Lack of coordination	Ample coordination	Efficient	Achieves Sustained Competitive Advantage	Accrue Firm Flexibility	Reduced Resource Wastage	Increased Resource Wastage	Improved Work Productivity	Decreased Work Productivity
Change in Govt. Policies	Pearson Correlation	0.078	0.346	-0.076	0.160	0.160	0.039	-0.134	-0.107	0.047	1	.[a]	0.118	.593*	0.334	0.034	0.327	0.304	-0.245	-0.212	-0.010	-0.224	-0.164	0.024	-0.126	0.080	-0.086	0.159
	Sig. (2-tailed)	0.743	0.136	0.750	0.502	0.500	0.871	0.573	0.654	0.844			0.621	0.006	0.150	0.887	0.159	0.193	0.297	0.371	0.966	0.343	0.489	0.919	0.596	0.737	0.717	0.502
	N	20	20	20	20	20	20	20	20	20	20	20	20	20	20	20	20	20	20	20	20	20	20	20	20	20	20	20
Decreased Work Productivity	Pearson Correlation	0.402	0.133	-0.231	0.339	-0.041	-0.259	0.103	-0.315	-.456*	0.159	.[a]	0.292	0.149	0.088	-0.111	0.248	0.282	-0.117	.697*	-0.397	0.157	-0.217	0.056	0.291	.740*	0.032	1
	Sig. (2-tailed)	0.079	0.576	0.327	0.143	0.864	0.271	0.665	0.176	0.044	0.502		0.211	0.531	0.712	0.642	0.291	0.229	0.624	0.001	0.083	0.510	0.358	0.814	0.213	0.000	0.893	
	N	20	20	20	20	20	20	20	20	20	20	20	20	20	20	20	20	20	20	20	20	20	20	20	20	20	20	20
Efficient	Pearson Correlation	-0.125	-0.167	0.237	0.029	-0.025	0.210	0.363	.484*	0.117	-0.224	.[a]	-0.044	0.083	.500*	0.103	-0.215	0.026	0.011	0.096	0.281	1	0.167	0.151	0.140	0.115	.507*	0.157
	Sig. (2-tailed)	0.600	0.480	0.315	0.903	0.915	0.375	0.116	0.031	0.622	0.343		0.855	0.729	0.025	0.666	0.363	0.913	0.963	0.688	0.230		0.482	0.525	0.555	0.630	0.022	0.510
	N	20	20	20	20	20	20	20	20	20	20	20	20	20	20	20	20	20	20	20	20	20	20	20	20	20	20	20
Improved Work Productivity	Pearson Correlation	-0.137	-0.279	0.350	-0.040	0.171	.473*	0.273	.481*	0.050	-0.086	.[a]	-0.057	0.093	.483*	0.399	-0.226	0.007	-0.092	0.039	.461*	.507*	.473*	-0.092	0.366	-0.034	1	0.032
	Sig. (2-tailed)	0.566	0.234	0.130	0.868	0.472	0.035	0.244	0.032	0.833	0.717		0.811	0.695	0.031	0.081	0.338	0.977	0.701	0.871	0.041	0.022	0.035	0.699	0.112	0.887		0.893
	N	20	20	20	20	20	20	20	20	20	20	20	20	20	20	20	20	20	20	20	20	20	20	20	20	20	20	20
Resource deployment	Pearson Correlation	0.157	0.115	0.174	0.015	0.142	0.336	0.179	.499*	0.147	0.334	.[a]	-0.079	.549*	1	0.372	-0.255	0.140	-0.199	-0.126	0.080	.500*	0.346	0.212	0.126	-0.107	.483*	0.088
	Sig. (2-tailed)	0.509	0.628	0.463	0.949	0.549	0.147	0.450	0.025	0.537	0.150		0.741	0.012		0.107	0.278	0.555	0.400	0.595	0.736	0.025	0.135	0.371	0.596	0.653	0.031	0.712
	N	20	20	20	20	20	20	20	20	20	20	20	20	20	20	20	20	20	20	20	20	20	20	20	20	20	20	20
Work Arrangement	Pearson Correlation	0.000	0.345	0.208	0.240	-0.195	0.425	0.184	0.293	0.158	.593*	.[a]	0.287	1	.549*	.452*	0.349	0.128	-0.248	-0.031	0.188	0.083	0.133	-0.123	0.115	0.076	0.093	0.149
	Sig. (2-tailed)	1.000	0.137	0.378	0.309	0.410	0.062	0.437	0.211	0.506	0.006		0.219		0.012	0.046	0.132	0.590	0.291	0.898	0.427	0.729	0.577	0.607	0.628	0.750	0.695	0.531
	N	20	20	20	20	20	20	20	20	20	20	20	20	20	20	20	20	20	20	20	20	20	20	20	20	20	20	20
Enables Resource Control	Pearson Correlation	-0.263	-0.129	0.135	-0.067	-0.011	.798*	0.423	0.437	0.162	0.034	.[a]	-0.091	.452*	0.372	1	-0.009	-0.201	0.066	-0.028	0.279	0.103	.560*	-0.189	0.238	-0.105	0.399	-0.111
	Sig. (2-tailed)	0.262	0.589	0.572	0.779	0.963	0.000	0.063	0.054	0.496	0.887		0.703	0.046	0.107		0.971	0.395	0.781	0.906	0.234	0.666	0.010	0.424	0.311	0.659	0.081	0.642
	N	20	20	20	20	20	20	20	20	20	20	20	20	20	20	20	20	20	20	20	20	20	20	20	20	20	20	20

*. Correlation is significant at the 0.05 level (2-tailed).

a. Cannot be computed because at least one of the variables is constant.

Fig. 7.15 SPSS correlation results with respect to the seven factors

7.4.2 Deductions of SPSS Correlation Results

Based on Fig. 7.15, Fig. 7.16 was constructed to highlight factors with strong correlations for better visualisation and comparison.

As seen from Fig. 7.16, assigned factors with stronger correlations than other factors and statistical significance at the 0.05 level (two-tailed test) are boxed in red as follows:

1. Decreased Work Productivity highly correlates with Lack of Coordination;
2. Decreased Work Productivity highly correlates with Increased Resource Wastage;
3. Work Arrangement highly correlates with Change in Government Policies;
4. Enables Resource Control highly correlates with High Construction Turnover.

Therefore, these strong correlations indicate strong relationships between these factors, which could be possible key reasons for the prevalent use of subcontracting practices, how subcontracting is being conducted, and the extent of influence these assigned factors have on construction productivity. In addition, two of the factors seem to have a rather significant influence on most of the other factors, as they have moderate correlation levels and are also statistically significant at the 0.05 level (two-tailed test). The strong relationship displayed by these two factors with other closely related factors is highlighted in blue, as shown in Fig. 7.16. The two factors have been identified as Improved Work Productivity and Resource Deployment. Hence, these two factors have a moderately strong correlation and are statistically significant, which serves to explain the relationship between the other assigned factors, which will also contribute more in enabling a plausible explanation for the research problem.

7.5 Summary of Results Analysis

Based on the analysis results from earlier sections, a summary of the key results derived from various methods of content analysis are tabulated in Table 7.3. Subsequently, logical inferences and deductions will be made based on the various key results. Thereafter, quotes will be extracted from interviewees' transcripts to further substantiate the analysis findings as well as their respective inferences and deductions. These are all shown in Table 7.3.

7.6 Conclusion of Analysis Results

The following are the conclusions from the results:

1. Resource Deployment is the main driving force behind the prevalence of subcontracting practices.

Chosen Factors with strong correlations

		High construction turnover	Large scale work	Economic Downturn	Change in Govt. Policies	Work Arrangement	Resource deployment	Lack of coordination	Ample coordination	Efficient	Achieves Sustained Competitve Advantage	Increased Resource Wastage	Improved Work Productivity
Decreased Work Productivity	Pearson Correlation			-.456*				.697*				.740*	
	Sig. (2-tailed)			0.044				0.001				0.000	
	N			20				20				20	
Improved Work Productivity	Pearson Correlation	.473*	.481*				.483*		.461*	.507*	.473*		
	Sig. (2-tailed)	0.035	0.032				0.031		0.041	0.022	0.035		
	N	20	20				20		20	20	20		
Resource deployment	Pearson Correlation		.499*			.549*				.500*			.483*
	Sig. (2-tailed)		0.025			0.012				0.025			0.031
	N		20			20				20			20
Work Arrangement	Pearson Correlation				.593*								
	Sig. (2-tailed)				0.006								
	N				20								
Enables Resource Control	Pearson Correlation	.798*				.452*					.560*		
	Sig. (2-tailed)	0.000				0.046					0.010		
	N	20				20					20		

*. Correlation is significant at the 0.05 level (2-tailed).

a. Cannot be computed because at least one of the variables is constant.

Fig. 7.16 Factors with strong correlations

Table 7.3 Summary of results analysis

S/N	Forms of analysis	Key results	Inferences/deductions	Conclusion
1.	Frequency—word clouds	With reference to Fig. 7.1: • The largest word clouds are Resource Deployment in orange and Improved Work Productivity in yellow • Other notable word clouds are Work Arrangement, Decreased Work Productivity, Increased Resource Wastage, Efficient, and Enables Resource Control, which are circled in blue	• The two factors Resource Deployment and Improved Work Productivity have a higher frequency of occurrence than any other assigned factors	A priority for further analysis is to determine whether there would be a possibility of a relationship between these two factors and their significance to other factors as well
2.	Frequency—table form	With reference to Fig. 7.2: • The frequency counts for both factors Resource Deployment and Improved Work Productivity stand at 64 and 58, respectively. In addition, these two assigned factors have appeared in all 20 transcripts	• This result further substantiates and concurs with the results from the word clouds	This is a significant result, requiring further analysis
3.	Frequency—horizontal bar chart	With reference to Fig. 7.3: • The highest code frequency is Resource Deployment, followed by Improved Work Productivity • Other significant factors are Work Arrangement, Decreased Work Productivity, Increased Resource Wastage, Efficient, and Enables Resource Control • There is no bar chart for the code Globalisation as no value is attributed to its frequency in Fig. 7.2	• All 20 interviewees did not view the effect of globalisation as an immediate threat to their livelihood and industry • It is plausible that some interviewees had little or no knowledge on the topic of globalisation	This concurs with previous results and requires further analysis
4.	Frequency—pie chart	With reference to Fig. 7.4: The frequency distribution of the prominent factors are as follows: • Resource Deployment—10.9% • Improved Work Productivity—9.9% • Work Arrangement—8.9% • Decreased Work Productivity—7.8% • Efficient—7.3%	• The factors Resource Deployment and Improved Work Productivity are the top two factors that have a substantial proportion of occurrence throughout the 20 transcripts	There is a need to analyse the relationship, if any, between the two factors as well as their relationships with other factors to determine their importance

(continued)

Table 7.3 (continued)

S/N	Forms of analysis	Key results	Inferences/deductions	Conclusion
		• Enables Resource Control—7.2% • Change in Government Policies—4.8% • Ample Coordination—4.8%		
5.	Cluster analysis—similarity index diagram	With reference to Fig. 7.8: • The similarity index between factors Resource Deployment and Improved Work Productivity is 1	• There is evidence of extremely strong similarity between these two factors and hence a strong relationship between these two factors, where the occurrence of one will affect the other	This proves that a strong relationship exists between the factors Resource Deployment and Improved Work Productivity. This explains how the occurrence of Resource Deployment will lead to Improved Work Productivity
6.	Cluster analysis—dendrogram results and agglomeration table screenshot	With reference to Fig. 7.7: • Factors that are in Cluster 1 (red) have more prominence and similarity than other clusters, as Cluster 1 has the largest number of factors and the most factors with a higher similarity index of close to 1 With reference to Fig. 7.9: • The agglomeration table lists factors that have the highest similarity index, also known as relationships of similarity, between factors. The top six nodes are chosen, as they have a high similarity index of 0.75–1.00 The top six nodes comprise the following seven assigned factors: • Resource Deployment Improved Work Productivity • Work Arrangement • Decreased Work Productivity • Efficient • Enables Resource Control • Change in Government Policies	• The seven chosen factors are more feasible in determining the reasons contractors predominantly adopt subcontracting practices and the effect of such practices on construction productivity	The seven factors will aid in focusing on the key factors or reasons that contribute to the prevalence of subcontracting practices and their associated effect on construction productivity. This enables the ease of further analysis

(continued)

Table 7.3 (continued)

S/N	Forms of analysis	Key results	Inferences/deductions	Conclusion
7.	Cluster analysis — co-occurrences table and tabulated results table	With reference to Table 7.2 and Fig. 7.11: The factors Resource Deployment and Improved Work Productivity have a co-occurrence frequency of 20 Other prominent co-occurrence values associated with the other chosen factors are as follows: • Resource Deployment versus Change in Government Policies—17 • Resource Deployment versus Decreased Work Productivity—18 • Resource Deployment versus Efficient—18 • Improved Work Productivity versus Change in Government Policies—17 • Improved Work Productivity versus Efficient—18 • Work Arrangement versus Efficient—17 Work Arrangement versus Improved Work Productivity—18	The presence of close proximity or relationships between these two factors is indicated, as they both keep occurring together for a significant 20 times	Further evidence of closeness of relationships strengthens the initial deduction that there is a cause and effect relationship between the factors, which indicates Resource Deployment will lead to Improved Work Productivity Other prominent factors with high co-occurrence values are secondary reasons for the prevalence of subcontracting practices and the effect on construction productivity
8.	Cluster analysis —proximity	With reference to Fig. 7.12: • Chosen code Resource Deployment and the given code Improved Work Productivity have the highest proximity value at 1.0. Moreover, both factors have the highest instances of co-occurrences at 20. With reference to Fig. 7.16: • This depicts a tabulation of the key results from the proximity analysis, where the two most notable given factors are Resource Deployment and Improved Work Productivity • However, the proximity value for selected code Decreased Work Productivity and given code Resource Deployment has a high value of 0.9	• These two factors have very strong proximity or share an extremely close relationship with each other, as they always co-occur • The two factors are the most significant among all other assigned factors • The factors Decreased Work Productivity and Resource Deployment also share a close relationship due to their high proximity value. This points to the possibility that	There is a high degree of possibility that the key reason for the prevalence of the subcontracting practices in the local industry is due to the flexibility given to firms via Resource Deployment. This enables firms to achieve an outcome of Improved Work Productivity

(continued)

Table 7.3 (continued)

S/N	Forms of analysis	Key results	Inferences/deductions	Conclusion
			the negative trend of a drop in productivity might occur if resource deployment is not implemented properly	
9.	Frequency Matrix Analysis	With reference to Fig. 7.13: • Code Resource Deployment (B) follows Work Arrangement (A) 28 times, which is the highest frequency of one code following another • A notable low frequency value for code Change in Government Policies occurs at 12 Other prominent frequency sequences among factors are as follows: 1. Efficient typically follows Resource Deployment, which stands at a frequency of 25. This means that the type of resources that are deployed at the work site by contractors will affect the efficiency of the job 2. Enables Resource Control mostly follows Resource Deployment, which stands at a frequency of 22. This means that the way in which contractor firms deploy their resources will help them accrue savings if Resource Deployment is carried out properly 3. Improved Work Productivity mostly follows Efficient, which stands at a frequency of 20. This means that, when contractor firms work efficiently, their construction work productivity will increase as well 4. Improved Work Productivity mostly follows Resource Deployment, which stands at a frequency of 19. This means that the way in which contractor firms deploy their firm's resources will cause their construction work productivity to increase as well 5. Decreased Work Productivity mostly follows Increased Resource Wastage which stands at a frequency of 19. This means that when there is a rise in resource wastage on site, construction work productivity for firms will drop	• The contractor firms typically discuss and agree on certain work arrangements prior to deploying their firm's resources where, based on the agreed work arrangements, the firms would then know which and how much resources to deploy • Depending on how efficient the resource deployment is, there could be an increase in resource wastage, which contributes to a decrease in work productivity • In contrast, if Resource Deployment is done efficiently, not only will firms accrue savings due to the ability to control resources well but also this will ultimately enable an increase in construction productivity	Other secondary factors and their respective frequency matrices point to several other cause and effect relationships, suggesting how some of the seven chosen factors and their inter-relationships among each other will lead to the key relationship of Resource Deployment that contributes to Improved Work Productivity Change in Government Policies apparently does not have much influence on Resource Deployment by contractor firms

(continued)

Table 7.3 (continued)

S/N	Forms of analysis	Key results	Inferences/deductions	Conclusion
			• Unfortunately, the low frequency result indicates that code Change in Government Policies does not have much influence on how the firms deploy their resources	
10.	Pearson's correlation analysis	With reference to Fig. 7.15: • The seven chosen factors with respect to other factors and their correlation values are all statistically significant (two-tailed test) and have moderate to strong levels of correlation The following chosen factors have higher correlation values at the 0.05 level of significance and have been highlighted in green and boxed in red as follows: 1. Decreased Work Productivity highly correlates with Lack of Coordination 2. Decreased Work Productivity highly correlates with Increased Resource Wastage 3. Work Arrangement highly correlates with Change in Government Policies 4. Enables Resource Control highly correlates with High Construction Turnover Other remaining chosen factors that have moderate correlation values and are statistically significant (two-tailed test) at the 0.05 level have been highlighted in green in Fig. 7.15 • The most distinct observation can be made on the factors Improved Work Productivity and Resource Deployment, as highlighted in blue in Fig. 7.15. These two factors have a significant influence on most other factors due to their moderate correlation levels and statistical significance at the 0.05 level (two-tailed test)	The strong correlations indicate strong relationships between these factors, which are possible key reasons for the prevalent use of subcontracting practices, how subcontracting is being conducted, and the extent of influence these assigned factors have on construction productivity The factors Resource Deployment and Improved Work Productivity are the main factors that contribute to a cause and effect relationship, where contractor firms that embark on Resource Deployment as a strategy for subcontracting, given its flexibility, will likely be able to achieve an increase in construction productivity	As the p-value from the correlation analysis also ascertains the validity of the relationship, it can be concluded that code Resource Deployment actually is the main driving force behind the prevalence of subcontracting practices As such, Resource Deployment by firms will most likely lead to the outcome of Improved Work Productivity Nonetheless, there are also other factors that are secondary reasons to support the occurrence of such a positive trend

2. Resource Deployment by firms will most likely lead to the outcome of Improved Work Productivity, a positive trend occurrence.
3. The chosen factors that are secondary reasons to support the occurrence of such a positive trend are as follows:
   - Efficient
   - Enables Resource Control
   - Work Arrangement
4. However, there is evidence to suggest that there are other factors that can influence the code Resource Deployment, resulting in a negative trend of Decreased Work Productivity, although such a possibility is not high. With regards to the proximity analysis, Decreased Work Productivity and Resource Deployment also share a close relationship due to their high proximity value of 0.9.
5. The following factors, although not deemed to be significant results during the analysis, can still attribute to a decline in work productivity: Lack of Coordination and Increased Resource Wastage.

Explanation of the above five points

The results from the analysis determined that the factor Resource Deployment is the main reason for contractor firms to adopt subcontracting practices, which results in its widespread adoption. This is attributed to its flexibility in allowing the firm's resources to be deployed under various changing circumstances among particular projects undertaken by the firm, which minimises the firm's risks. Thus, this enables firms to be more prepared for uncertainties. As such, any phrases, sentences, or words that bear similar meaning in the 20 transcripts to the code Resource Deployment are tagged to this code during the coding stages of the content analysis. This coding sequence is the same for all other factors as well.

Other factors, such as Enables Resource Control and Work Arrangement, are supporting reasons for contractor firms to adopt subcontracting practices. Basically, such factors suggest that this aids firms in accruing savings in resources, reducing costs via resource control, and enabling flexibility in deciding how best to divide the workload with other contractors in their arrangement, while still completing the work. As such, the key benefit of subcontracting practices is primarily for its efficiency, which is sought after by contractors, as shown.

However, despite the strong positive trend of resource deployment that can ultimately contribute to improved construction productivity, care must be taken regarding the implementation of resource deployment. If not carried out properly, the presence of the factors Lack of Coordination and Increased Resource Wastage will occur, which threatens to cause a decline in construction productivity.

References

Aldenderfer, M. S., & Blashfield, R. K. (1984). *Cluster analysis*. Thousand Oaks: Sage.

Anderberg, M. R. (1973). *Cluster analysis for applications*. New York: Academic.

Boost Labs. (2014). *What are word clouds? The value of simple visualizations*. Retrieved 28 August 2016, from http://www.boostlabs.com/what-are-word-clouds-value-simple-visualizations/.

Guest, G., & McLellan, E. (2003). Distinguishing the trees from the forest: Applying cluster analysis to thematic qualitative data. *Field Methods, 15*(2), 186–201.

Hubalek, Z. (1982). Coefficients of association and similarity, based on binary (presence absence) data: an evaluation. *Biological Reviews, 57*(4), 669–689.

LeCompte, M. D., & Schensul, J. J. (1999). Analyzing and interpreting ethnographic data. In J. J. Schensul & M. D. LeCompte (Eds.), *Book five of the ethnographer's toolkit*. Walnut Creek: Altamira Press, a division of Sage Publications.

Lugwig, J. A., & Reynold, J. F. (1988). *Statistical ecology: A primer on methods and computing*. New York: Wiley.

MacQueen, K. M. (2005). What to look for in software for qualitative data analysis. In P. R. Ulin, E. T. Robinson, & E. Tolley (Eds.), *Qualitative methods in public health: A field guide for applied research* (pp. 172–174). San Francisco: Jossey-Bass.

Niwattanakul, S., Singthongchai, J., Naenudorn, E., & Wanapu, S. (2013). Using of Jaccard coefficient for keywords similarity. In *Proceedings of the international multiconference of engineers and computer scientists 2013* (Vol. I, pp. 380–384), IMECS, March 13–15, Hong Kong.

Oxford Dictionary. (2016). *Word cloud—definition of word cloud in English*. Retrieved 28 August 2016, from http://www.oxforddictionaries.com/definition/english/word-cloud.

Partek, (2012). *Partek user's guide: Hierarchical clustering analysis*. Chesterfield, MO: Partek Incorporated.

StatsDirect. (2016). *Simple Linear Regression and Correlation. Statsdirect.com*. Retrieved 28 August 2016, from http://www.statsdirect.com/help/default.htm#regression_and_correlation/simple_linear.htm.

Wheaton College. (2016). *How to read a dendrogram*. (2016). *Wheaton College LEXOMICS*. Retrieved 28 August 2016, from http://lexomics.wheatoncollege.edu.

Chapter 8
Interview Findings

Abstract This chapter establishes the link between industry fragmentation due to the adoption of subcontracting practices and its effects on productivity based on an analysis of the fieldwork results. This chapter presents verbatim quotations from the interview transcripts to explain the occurrence and meaning of the analysis results. The verbatim quotations from the interviewees further shed light on the various reasons why the multilayer subcontracting practices are prevalent in the construction industry despite its perceived limitations.

After the derivation of the statistical analysis results from the research findings, this section seeks to quote relevant sections from various transcripts to explain the occurrence and meaning of the analysis results. In addition, such quotes from various interviewees further shed light on reality by providing various reasons and comments regarding the derived results. Quotes will be extracted from the interview transcripts to support the results.

8.1 Resource Deployment is the Main Reason for the Prevalence of Subcontracting Practices

Through the process of data mining and analytics, the most prominent factor that suggests the reason for such widespread adoption of subcontracting is the factor Resource Deployment. As a main contractor from Interview 4 (I04) put it, in general the reasons for subcontracting are that: "*It will improve productivity, as these subcontractors are more familiar in their field of work. They have the capacity to complete that particular job*".

In addition, this view is further reinforced as a common viewpoint by many contractors. The main contractor from Interview 9 (I09) put it simply: "*All those specialised contractor, with regards to productivity, they are in this trade for many years already. Thus, they should have the skill and resources to do their trades*".

L. Sui Pheng and S. Jia Zheng, *Construction Productivity in the Multilayer Subcontracting System*, Management in the Built Environment,
https://doi.org/10.1007/978-981-13-1831-3_8

To reiterate, it is not just the main contractor that employs subcontracting. Even subcontractors do so. As subcontractor from Interview 19 (I19) put it:

> If they can't handover or they handover late despite me rushing my job within the much shorter timeframe but yet can finish the job for the main contractor, then it is not our problem. This is because you have to work day and night to meet the shorter deadline than usual. So this is the point that we look for sub-subcontractors. Mainly because of time constraints.

To ensure the success of the factor Resource Deployment, which will contribute to an increase in the firm's productivity, a main contractor from Interview 13 (I13) stressed that a key part is contingent upon the management of both main contractors and subcontractors:

> If productivity we want to maintain at a very high level, we must ensure that our own supervisors and foreman are all up to speed and they are very capable to push the works. So productivity doesn't rely on the workforce itself, it relies a lot on the management. If it is the subcontractor's management, they will definitely want to do things as fast as possible, do it the first time so that their cost can be as low as possible so that they can earn as much money as possible.

As such, the strategies employed in relation to factor Resource Deployment play an important role in ensuring productivity gains. As shared by a subcontractor from Interview 10 (I10) on his strategy:

> Our 'sub-subcons' are specialised in their field of work, and it will be more efficient for them to do the job as it can be done more quickly and with less manpower. In this industry, if you are the 'subcon', it is still best for you to specialise as you can pull your resources together to ensure that your resources are more fully utilised, have lesser wastage, and you can concentrate your efforts on making sure you excel in a particular trade. Once word gets out into the industry that you are good in a particular trade, you will end up receiving many jobs. That's how you survive as a 'subcon' in Singapore's cut throat industry.

Such a resource deployment strategy is commonly employed by other contractors as well. As mentioned by another subcontractor in Interview 18 (I18):

> The specialist contractors will do their specialist work as your firm does not have the necessary expertise to do the work. Hence, you 'sub-out' to them. These specialist contractors have the necessary manpower, skills and tools to do the job and thus, they can give you more competitive advantage.

Hence, a common strategy to deploy the firm's resources is to determine whether the firm has the necessary expertise to complete the job. If the firm lacks such expertise or is hard pressed for time, they will adopt subcontracting as a viable means to ensure work is completed without unnecessarily wasting the firm's resources. This will ultimately aid in productivity gains for such firms that are strongly desired.

8.2 Resource Deployment by Firms Will Most Likely Lead to the Positive Outcome of Improved Work Productivity

The main contractor from I13 discussed the merits of subcontracting, which enable firms to accrue productivity gains:

> Having subcontractors in the positive side is that they will enable higher productivity as they are specialised in a particular works and usually they want to get things done fastest, shortest, and easiest and cheapest way so that they can get as much profit as possible.

8.3 Secondary Reasons, Where the Practice of Subcontracting is Encouraged Due to the Factors Efficient, Enables Resource Control, and Work Arrangement, Which Contribute to Productivity Gains

The majority of the interviewees commented that such subcontracting practices aid in efficiency, where work is done within the shortest possible time. As summarised by a subcontractor from Interview 17 (I17):

> Usually, I go for efficiency in getting the job done. When you have your own workers, they will take their own time and do the works slower. However, for the 'subcons', once they finish their job, they get paid. Hence, more incentive for the 'subcons' to work faster.

In the case of subcontracting that enables control of resources by the firm, a main contractor from Interview 3 (I03) felt such practices ensured effective resource deployment:

> The same manpower, we can split up the work, rather than face with a situation where there is not enough workers and need to bring in more workers. We will discuss with the 'subcon' to limit the number of workers, where there is adequate manpower to do the job.

The effectiveness of resource deployment is even more crucial, especially in recent times, where the Singapore government has made changes to the existing labour policies. This has, no doubt, caused many firms to feel the brunt of manpower issues. Such issues are discussed in earlier chapters under the literature review. However, as seen from the results analysis, a change in government policies does not seem to affect the deployment of a firm's resources. This is validated by the main contractor from Interview 8 (I08):

> Due to the ability of our firm to tender for projects within the company resources so as to make full use of the company resources. Ultimately, it comes down to the market condition.

As such, this essentially indicates how a particular firm works within its means by ensuring the firm's resources are fully utilised at any given point in time, based on the current market conditions. Therefore, it might be somewhat unreasonable to attribute the difficulty in deploying a firm's resources to a change in government policies.

Moreover, the adoption of subcontracting practices can also be attributed to the flexibility that is allowed in various work arrangements for firms. Not only does it allow for the continuous flow of jobs to firms that is manageable but it also ensures their survival in a competitive industry. This is a view shared by a subcontractor from Interview 12 (I12):

> For my case, I am more versatile as I am still doing builder works where me and Mr X [main contractor] work together. If I get the entire work package, I will sub some of the works out to him or vice versa, a flexible arrangement.

This view concurs with another interviewee from a main contractor (I09):

> Sometimes, we also become a subcontractor when we have a special project. One example would be when we do a jet engine test facility for a client where they call tender for test engine cell. So the test cell alone which they want to build, they 'sub-out' to the mechanical engineering firms. So one of our Japanese counterpart firm, they got the job. However, some of the civil engineering part of the job, they sub it out to us. Like those structural works, architectural work and basic M&E works. Only the specialist equipment is installed by them. So they become the 'main con', we become the 'subcon'. This is a very special arrangement.

8.4 Possibility of Negative Trend of Decreased Work Productivity Due to Resource Deployment Strategies is not Highly Significant

This can occur when contractors fall short in their planning and implementation Resource Deployment strategies for their various jobs, thereby causing ineffective allocation of various forms of resources. This eventually leads to a decrease in their firm's work productivity, as lamented by a subcontractor from Interview 5 (I05):

> You have your own workers doing a particular trade. Then that particular job is not available. So what do you do? If the core competency of your workers is to do duct fabrication and the job is not available to them, then in order to maximise your workers, you ask them to do duct cleaning as well. This actually happened to my core subcontractor (sub-subcontractor). Even though this is not the core strength of the workers, but he rather do this, as he don't want to give his workers idling. Keep them occupied and reduce his cost. However, this is not productive, as it is not using the correct worker for the correct job.

8.5 Lack of Coordination and Increased Resource Wastage Attributed to Decline in Work Productivity

Besides the ineffective deployment of resources that threatens to incur a productivity decline for firms, other common factors like Lack of Coordination and Increased Resource Wastage are contributing factors that can derail a firm's productivity gains. Such a conclusion is shared by both main contractors and subcontractors. Regarding the factor Lack of Coordination among parties, a main contractor from Interview 7 (I07) shared that:

> Cannot have too many layers of subcontracting. If too many layers then got problem already. For example, I am the main con and I 'sub-out' tiling works to Company A. Then, Company A might 'sub-out' to Companies C and D. Hence, got three to four layers. Then, at most, we try to choose direct sub, and we don't expect them to have a few more layers under them.

Similarly, on the issue of Lack of Coordination, a subcontractor from Interview 20 (I20) commented that:

> Most of the 'subcons' face the same problem is that they lack of planning before implementation. They tend to take too many jobs as they are specialist and not every job is the same and thus this is a problem. So lack of preparation or understand the contract of the scope of the works. So when they don't understand, they don't plan. So this results in full of surprises when they go to the site.

Another main contractor from Interview 6 (I06) went to the extent of giving a prominent example of such a lack of coordination among various parties on employing building information modelling (BIM) in the industry that is supposed to aid in productivity gains:

> This is why BIM might not be widely adopted because of coordination issues where your main contractor knows how to use BIM but not your subcontractors. As such, with regards to productivity improvement at the end of the day, it is at the 'main con's' expense whether to assist the 'subcons' to improve productivity, where there is a price to be paid when we try to optimise productivity and cost.

The other issue that plagues productivity gains is the factor Increased Resource Wastage. Market conditions and government policies have somewhat contributed to this factor. From time to time, this issue has been brought up by both main contractors and subcontractors. A subcontractor from Interview 15 (I15) complained about this issue:

> The MYE and safety requirements will incur all the cost… You need the risk assessment and safety supervisor to change one bulb. Then, the fellow who bring the ladder to climb, also need somebody to hold the ladder; that is the regulation. Then, the guy that climb also need a safety belt under the supervision of the safety supervisor, who just stand there and watch. So the three person cost you see how heavy, is it productive? In Europe and Australia, only one person do the work. They take care of themselves and they are the safety supervisor and everything they do themselves. So in fact, the three workers' salary add up together is equal to or even higher than one Caucasian worker. So it is not productive at all… They (foreign workers) actually are farmers, and they come here to become

workers. You ask them to do construction trades, they don't know. Hence, you must train them. Then, when you finish training them, they reach their maximum length of stay here and they have to go back already. Hence, we like a training centre for them. All these issues make us very hard.

Another subcontractor from Interview 19 (I19) concurs with the above mentioned view of manpower wastage as well that is plaguing almost all contractors:

Of course, when the new guy comes in, at least must go for one to two years of training to get the necessary skills, have to go for all these training. But important is the worker wants to come back and work for us again … and other companies just offer slightly more then take our former workers whom we train until everything ok already. Also, the worker already knows where he wants to go, so back to square one, have to conduct retraining again, then the workers leave again after a few years, then we retrain new batch again. A cycle you see. Every contractor also facing this kind of problem.

In addition, a main contractor from Interview 3 (I03) aptly stated how 'Lack of Coordination amongst parties' and 'Increased Research Wastage' can be attributed to the fragmentation of the industry, where the presence of multiple parties, resulting in a hierarchy that can easily lead to miscommunication, hinders the productivity gains:

The traditional design-bid-build got the clerk-of-work, register and technical officer, all doesn't work! They are just here to check paper work as to whether you got submit or not … They [client] got the RTO, RE [resident engineer] to assist all these three parties [architect, structural, and M&E], to look at the main contractor. Like this how to increase productivity? If the contractor understands what the owner wants, we build according to what the owner wants. We tell and bring the architect along and tell him that since he is under our payroll now, please address what the owner wants and we build together. Can you imagine the traditional way, how tedious it is? … Then under the owner, still got NSC, then they also have their direct 'subcon'. So can you see how many trades they have here?

Hence, the presence of fragmentation in the industry does have an overall effect on construction productivity, where, due to the multiple parties present within a single project, the need to allocate work according to their job scope gives rise to the prevalence of subcontracting practices. This is further substantiated by a subcontractor from Interview 5 (I05) that has elaborated in detail on fragmentation in the local industry:

From the perspective of the main contractors, they don't really want to care much. The 'main con' has so much problems handling the clients when they request for multiple changes. At the same time, the main contractor must relay the instructions passed down from the client's consultant on certain works that needs to be done in the building. Then the 'main con' must liaise with its 'subcon' and determine if their 'subcon' can do the job …

Note that project managers [PM] don't know the prices. It is the QS [quantity surveyor] that knows the price. PM only knows who do what, when are the 'subcons' coming, etc. PM is the guy that does the coordination, and that's the realm of the PM. 'Main con' don't like to bother with what the 'subcon' is doing but the 'main con' has a project manager, RTO [resident technical officer], RE [resident engineer], etc. So the RTO is hired by the main contractor, whose job is to make sure that the works are done properly. Besides the RTO, RE is under the client where their job is to check on the main contractor and ensure works are done properly in accordance to specifications and standards.

> RE's job is to check on the 'main con'. There are a few RTOs below the RE, namely, structural RTO, M&E RTO, P&S RTO, etc. The main contractor also got their own RTOs. Mind you that they are not friends as the main contractor's side is to make sure that the works are done properly and he signs off. Then, he brings the work to the client's RTO. Then, once all works in order after client's RTO check, then, it will be sent to client's RE for final checks. Once client's RE signs off, then, the job will keep moving. RE's job is just to represent the client.

In conclusion, the extraction of the various quotes from different interviewees has provided much insight on reasons that serve to explain the analysis results. Such a process will enable readers and future researchers to appreciate how the formation of the industry structure is based predominantly on MLSS, the motivations as to why subcontracting practices are executed in the way in which they are, and how subcontracting practices will aid in productivity gains when done appropriately.

Chapter 9
Conclusion

Abstract This chapter provides the recommendations, highlights the limitations in the research process, concludes the study, and suggests possible areas for further research. This study has made some useful academic contributions, as it seeks to provide the academic community with the necessary insight into the underlying reasons and motivations that drive construction firms to behave the way they do and how they operate under trying circumstances given the challenging market conditions. It offers a comprehensive understanding of the state of the construction industry and a more in-depth view of the intricacies of subcontracting practices adopted. In addition, it also seeks to determine qualitatively how these subcontracting practices, when undertaken in Singapore, help in achieving productivity gains in the context of organizational learning.

The purpose of this study is to explore how fragmentation of large and small firms in Singapore affects construction productivity when operating under the MLSS and the reasons for the adoption of subcontracting practices by local contractors.

9.1 Review of Research Aim and Objectives

Through the literature review, the research gap was discovered, where there is still a lack of thorough understanding of the MLSS in the local construction industry as to how such fragmentation, deemed as the root cause of subcontracting, has an effect on construction productivity in Singapore. By investigating further, the research aim, its objectives, and the hypothesis have been achieved. These will be covered in detail in the next section. A recap follows.

Research Aim: To explore and better understand how fragmentation of large and small firms in Singapore affects construction productivity when operating under the commonly adopted MLSS in the context of the local construction industry.

L. Sui Pheng and S. Jia Zheng, *Construction Productivity in the Multilayer Subcontracting System*, Management in the Built Environment,
https://doi.org/10.1007/978-981-13-1831-3_9

Fulfilment of research objectives:

1. The construction industry is labour intensive by nature, which is common globally, as it consists of workers of varying skill levels, specialties, and technical and managerial positions. This has caused the prevalence of outsourcing globally, due to the popularity of specialisation. This is by no means different for Singapore's industry.
2. Singapore's construction industry is extremely fragmented. An analysis of a survey done by Singapore's BCA shows 16,823 firms for such a small industry, with the majority being small firms that stood at a whopping 16,139 firms in 2013.
3. By analysing the data from the BCA, the large presence of small and medium firms that have increased over the years is proof that fragmentation still firmly exists within the local industry. Therefore, it can be deduced that subcontracting practices are still prevalent within the industry, where it is a norm to subcontract up to 70% of the work.
4. It can be concluded with a high degree of certainty that subcontracting practices, if executed appropriately, can aid in the improvement of construction productivity in the industry.

The next section will recap the key findings of this study and the various methodologies that are adopted.

9.2 Summary of Study

The study began by establishing the research aim, its objectives, and the scope. The main crux of this study is concerning how and why current subcontracting practices in the local industry are done as well as their eventual effects on Singapore's construction productivity. In addition, three common classifications of contractors were identified: main contractors, subcontractors, and secondary subcontractors. These three classifications essentially help explain the formation of the commonly adopted MLSS in the local context.

With the adoption of the MLSS, another key issue of the construction industry emerges—fragmentation. Such an issue is not unique to Singapore; it happens globally to other construction industries of other countries as well. The basis of fragmentation is due to the prevalence of outsourcing, where it is attributed to the popularity of specialisation. As such, this gives rise to two main types of subcontracting practices: specialisation-oriented subcontracting and capacity-oriented subcontracting. In Singapore, these two types of subcontracting practices are commonly adopted, thereby enabling the existence of the MLSS. However, more often than not, the downsides and controversies surrounding the MLSS were revealed in the literature. Such downsides include reduction of profit margins and no value-add to the overall MLSS, while controversies, such as instances of poor project quality, budget overruns, and poor communication among parties, occur from time to time in other countries. Nonetheless, the MLSS can be successful if implemented properly, as seen in the case of Japan's construction industry. Therefore, it is with hope that, in the near future, Singapore's industry can take a step in this direction despite adopting the MLSS structure.

In the local context, fragmentation within the industry occurs in the form of the existence of many construction firms. As of 2013, there were about 16,823 firms in the local industry. Of which, 16,139 firms are small firms, employing less than 100 employees, while the remaining 684 firms are medium and large firms, which shows the extent of fragmentation due to the large presence of small and medium firms that has increased over the years. This is proof that fragmentation still firmly exists within the industry. Hence, it can be inferred that subcontracting practices are still prevalent within the industry today.

This is followed by the use of various management theories via readings of journals and books that seek to explain the adopted practices and their effects, based on the current circumstances of Singapore's construction industry. Three main theories that explain how and why contracting firms behave and operate in the MLSS framework were discussed. This began with the RBV approach, followed by Porter's (1980) Five Forces Model, and eventually ending with the flexible firm model. Throughout the course of the discussion, it is discovered that the flexible firm model is most suited to explaining the prevalence of subcontracting practices adopted in Singapore, as its theory and explanation sync well with current market conditions like globalisation and the new labour restructuring policies that came into effect in 2011. A combinative explanatory framework is then developed that serves to merge the three theories into a common platform that will serve to exemplify the reasons for the prevalence of subcontracting, which leads to the presence of MLSS due to their relative importance.

A crucial part of this study is the research design and analysis, where primary and secondary data collection from interviews and questionnaires were analysed as part of a research framework that consists of three stages: the identification stage, analysis stage, and findings stage. A total of 20 interviews were made, 10 of which were targeted at main contractors, while the remaining were for subcontractors. All interviews were transcribed and input into a data-mining and analytics software called QDA Miner 4.0. With the aid of a coding framework, the results of the data analysis have been achieved. From the results analysis, it can be concluded that the key reason for the prevalence of subcontracting practices is mostly due to the factor Resource Deployment, where it will most likely lead to the outcome of Increased Work Productivity, therefore contributing to the industry's overall rise in productivity. Other supporting factors for such a positive trend include Efficient, Enables Resource Control, and Work Arrangement. However, it is still possible that the factor Resource Deployment could have a tendency to cause a decline in work productivity should the presence of factors such as Lack of Coordination and Increased Resource Wastage occur.

In conclusion, the study has fulfilled its research aim and objectives and has shown that the action of outsourcing or subcontracting several workloads out to other parties would not necessarily lead to a decline in construction productivity. Therefore, the research hypothesis formulated for this study appears to be supported.

9.3 Implications of Study

This study has made some useful academic contributions, as it seeks to provide the academic community with the necessary insight into the underlying reasons and motivations that drive construction firms to behave the way they do and how they operate under trying circumstances given the challenging market conditions at present.

Moreover, it offers a comprehensive understanding of the state of the local construction industry and a more in-depth view of the intricacies of subcontracting practices adopted in the local context via the rich data collection from 20 experienced industry professionals that will be beneficial for further analysis. In addition, it also seeks to determine qualitatively how these subcontracting practices, when done in the local context, help in achieving productivity gains.

9.4 Limitations of the Study

Like all research studies, there are also some limitations with this paper. These limitations are listed below:

- The limiting factor is mainly time. Due to the time limitation of approximately one year, it is challenging to secure as many interviews as necessary to enable a higher accuracy of the results, even though this is desired. Hence, this study has proceeded with 20 interviews.
- With time constraints, validation of the results in the form of validation interviews with a few other prominent industry players cannot be done. Nonetheless, the utmost due diligence and care have been exercised when conducting the 20 interviews in order to extract as much information from the interviewees. In addition, multiple attempts have been conducted on data mining and analytics using the software during the analysis stage to minimise the occurrence of errors and ensure the consistency of the results.
- Content analysis, despite its efficiency and usefulness in qualitative analysis, also has weaknesses (Morris, 1996). They are listed as follows:
  - The software's inability to recognise the communicative intent (e.g., irony and sarcasm) of word usage (Krippendorf, 1980);
  - The inability of the researcher to provide an exhaustive list of every related word to a code or element;

- The inability of the software to resolve references to related pronouns appearing elsewhere in the text (i.e., that, them, etc.; D'Aveni & Macmillan, 1990);
- Human judgement is still required to reduce the risks of abstracted empiricism (Gephart & Wolfe, 1989) and to establish the validity and reliability of the computerised content analysis;
- The context in a content analysis is usually not considered or is highly constrained, thereby limiting the richness of the data results.

Despite these limitations, the study was able to establish a high degree of fragmentation in Singapore's construction industry to determine that the key reason for the prevalence of subcontracting practices is mainly due to the flexibility for construction firms to deploy their resources quickly and effectively and that subcontracting can eventually have a positive effect on Singapore's construction productivity. Moreover, this study has opened up new realms of topics that can be explored in future research based on the rich qualitative data that have been uncovered from the interviews.

9.5 Recommendations for Future Research

Based on the results analysis from the process of data mining and analytics, this research opened up other issues that have been plaguing the local construction industry for a very long time that warrant further investigation and research studies.

Issues such as industry irregularities, where main contractors commonly hold back progress payments to their subcontractors and exploit loopholes in the Security of Payment Act (SOP Act), are genuine concerns that are raised by subcontractors, as they have been unfairly treated and exploited. Although this happens from time to time, such events still have the potential to derail the industry's overall construction productivity. Therefore, it would be worthwhile to explore such issues in the near future.

In addition, to fully harness the efficiency and effectiveness of data mining and analytics, other forms of similar software can be adopted. For instance, the use of a text-mining software called WordStat can help filter through more interview data, thus enabling faster extraction of themes and trends, or it can be paired with other state-of-the-art quantitative content analysis tools to provide a much more refined analysis result.

9.6 Organizational Learning: The Way Forward

Organizational learning (OL) has emerged as one of the most crucial and sustainable competitive advantages in enhancing productivity (Guthrie, 2005; Levitt & March, 1988), leading to the growth of, for example, construction companies

(Barlow & Jashapara, 1998). OL began with workers or units learning from their own experiences, or from experiences of other units within the same organization (Argote & Ophir, 2002). Today, the fast pace of technological advancement and the rising expectations for innovations and improvements have made it difficult for organizations to rely solely on learning from within their organizations. This has led to many industries adopting processes by which they can learn from other organizations. Through the sharing of knowledge, companies and industry players can learn from within and among organizations in the dynamic and ever-changing industry. In fact, the construction industry is quite active in terms of learning and adapting management concepts and thinking from other sectors—such as lean concepts, Just-in-time principles, and value engineering. Moving forward, further study should similarly investigate the concept of OL in the construction industry, and especially focuses on the relationship between OL and productivity. The main challenge that the construction industry constantly faces is the fragmented nature of the industry: different industry professionals and trade players come in at different stages of the building delivery process, leading to ineffective coordination and poor communication. Moreover, every construction project is individually unique, and so it is challenging to capture the desirable portions of one particular project and to put these into practice for future projects. In addition, construction projects are generally complex and of relatively long durations—making documentation even more difficult.

OL is a relatively new term in the Singapore construction industry. Knowledge and human capital are the two most important resources for construction companies in resource-scarce Singapore, and OL provides the integrative solution for incorporating both of them together. This inevitably contributes to the relatively gentle learning curve found in the construction industry as compared to other industries in Singapore, leading in turn to low productivity rates. To increase Singapore's construction productivity (CP), it is essential to establish a steeper learning curve for companies. OL will enable companies and contractors to document construction projects and to store and share valuable experiences. Through proper documentation and sharing, companies and contractors can learn from past projects and apply useful lessons to future projects. This will then lead to a steeper learning curve than exists presently.

Organizational learning (OL) has become increasingly popular in many organizations. Although it suffers from a lack of clear definition (Hong, 1999; Örtenblad, 2001), most agree that OL is about acquiring knowledge (Curado, 2006a; Huber, 1991) and putting this knowledge to future uses, for the purpose of solving problems or to perform better through lessons learned. To distinguish from another term—the learning organization—which was sometimes used interchangeably with OL, efforts were made by Örtenblad, 2001), who defined the learning organization as a form of organization, while the latter (organizational learning) is an activity or processes of learning in an organisation. Learning organizations require effort to develop, while organizational learning exists without such efforts (Örtenblad, 2001). Broadly, as observed by Hong (1999), research into OL attempts to answer two questions—"how does the organisation learn" (Tsang,

1997), and "what has the organization learned" (Argyris & Schön, 1978). In terms of the "how" question, we can say that OL normally takes place on two levels—that of the individual and that of the organization. The main players that contribute to OL are the individuals within the firm (Kim, 1998). Torbert (1972) defines individual learning as becoming aware of the qualities, patterns, and consequences of one's own experience as one experiences it. In addition, some argue that OL is not just a cumulative result of individual members' learning (Argirys & Schön, 1978; Curado, 2006b; Hedberg, 1981), but that even if the members (i.e., the leaders and employees) of the organization change, the organization's memories can still preserve certain behaviours, mental maps, norms, and values over time (Hedberg, 1981). With respect to the "what" question, the framework provided by Bell et al. (2002) encompasses four schools of thought—economic, developmental, managerial, and process-oriented—and is useful in understanding the context. According to Bell et al. (2002):

- The economic school of thought (learning by doing) is focused on encoding inferences and understanding that accrue from the repetitive action and subsequent reactions that occur with continuous production. What can be learned usually comes from the detection and correction of errors (Argyris, 1977).
- The developmental school of thought (learning by evolution) acknowledges that the learning organization represents a phase or objective in the evolution of its development. It describes the organizational development resulting from a recurrent and cumulative sequence of learning. Examples include structured trade-based apprenticeships and professional accreditation programmes (Tennant & Fernie, 2013).
- The managerial school of thought takes place as a result of management-led interventions and addresses higher-order learning. It happens as a form of a calculated response to changing circumstances (Tennant & Fernie, 2013). For example, workforce development is needed when there is a legislative requirement from government.
- The process school of thought represents all the learning processes that simulate knowledge creation, knowledge sharing, and coproduction (Hong, 1999; Tennant & Fernie, 2013). What is likely to be learned, in the construction context, can come from the platform of intranets, project planning and scheduling, and formal site meetings (Tennant & Fernie, 2013).

An active OL process is essential in companies to translate what has been learned into knowledge that can be stored and shared. If the firm fails to realize the importance of OL, what has been learned will just become a by-product of normal business operations (Tsang, 1999), and the advantages of OL will not have been fully exploited. It is thus imperative to employ OL in today's ever-changing environment in order to capture useful knowledge and to communicate it to the individuals in the organization, so that the capabilities of both the organization and the individuals can be built up (Kim, 1998). Knowledge of best practices can be put into use while potential pitfalls can be avoided. Through this, organizations will be able to maintain and improve their competitiveness, productivity, and innovation (Dodgson, 1993). Translating these promising benefits into the construction context involves cost deductions, improved quality and productivity, satisfied clients, and others.

In terms of hindrances, much has been said about the variety of factors that might affect the implementation of OL. For example, the quantity of resources allocated to OL plays a central role in affecting the overall process of OL. Generally, greater amounts of resources contribute to more effective and more efficient OL. Dodgson (1993) has discussed how the Research and Development (R&D) departments of organizations constitute major sources of learning, and are thus central to the continuing existence and prosperity of various activities; they thus lead in turn to more effective OL. Additionally, the budget that organizations set aside is another primary factor that can encourage or limit OL. Learning has always been a costly process. It is difficult to assign a financial value to individual learning and OL (Dodgson, 1993), considering that both actual costs and opportunity costs are incurred by OL. In addition, organizations will learn about new technological developments through their R&D departments. Technological innovations have led to rapid advancements in information technology (IT), which is essential in aiding OL. Costs are incurred in the OL process when transforming experience into routines, and when recording these routines (Levitt & March, 1988). Technology has also helped to reduce the costs of recording routines, and has allowed information to be shared and communicated more easily, more quickly, and more conveniently—though the initial cost of implementing such technology is high. Moreover, OL is dependent on external factors, which include prior conditions and the organizational environment. The way that an organization learns and reacts is influenced by its past and by its collective learning through individuals (Dodgson, 1993). Individual learning is dependent on the presence of existing knowledge, and on what was already known to members themselves and to other members of the organization. Sometimes, individuals may have created certain prior mental modes or mindsets, and these are not easy to change once they are embedded in those individuals. For OL to be optimal, there must be a balance between change and stability in the organizational environment. Too much stability within an organization can be dysfunctional, as people might not realise that there is a need to learn and change, with the result that processes and behaviours become obsolete (Fiol & Lyles, 1985). At the same time, too much change can make it difficult for learners to map their environment.

Much research has investigated OL and the resulting productivity enhancements, but not much light has been shed on the construction sector. At the other end, most research has discussed the importance of OL in the construction industry (Barlow & Jashapara, 1998; Kamara, Augenbroe, Anumba, & Carrillo, 2002) and the factors that affect OL (Egbu, 2004), but there is a lack of discussion bridging the two. For example, Barlow and Jashapara (1998) discussed the role of OL and inter-firm partnering in the UK construction industry, and the factors which influence the transfer of knowledge between organizations, but the impact on productivity was not touched on. Kamara et al. (2002) described the importance of Knowledge

Management (KM) in the construction sector, but did not link the management of knowledge to OL, nor did they elaborate on how the management of knowledge in the construction sector would bring about a competitive advantage in terms of enhanced productivity. Egbu (2004) identified some of the critical factors leading to successful KM practices and improved organizational innovations. However, critical factors leading to successful OL and CP were not discussed. Argote (2012) explored inter-OL by transferring knowledge from other organizations, and looked at how this improves the firm's productivity. Yet that review did not look into the sharing and communication of knowledge within the firm (intra-OL), and serves only to provide a generic viewpoint. It does not look into OL and productivity in the construction sector.

The culture of the Singapore construction industry is to move on to the next project after completion of the previous one, with little or no attention paid to reviewing the completed project. Valuable experience is not being documented, and hence precious knowledge is not being retained and transferred within the firm. Without knowledge repositories, OL is essentially impossible. Many bad practices or mistakes are repeatedly made and seen, due to the lack of documentation and the inability to learn from these mistakes. This leads to "lower than could have been" productivity, had the companies learned from their mistakes and avoided repetitions of errors. Seeing this, BCA has set up online sharing platforms, such as CORENET and Construction Infonet. The former is capable of providing a high level of automation and integration for many construction and development processes, and it is hoped that, with the use of information technology, high levels of efficiency, productivity, and quality can be achieved in construction projects (Sing & Zhong, 2001). CORENET acts as a central repository for building codes, regulations, and circulars published by the various building and construction regulatory agencies in Singapore through the e-Information System, as an online platform to facilitate the submission and processing of building-related applications and documents through the e-Submission System, and to provide easy access to the basic **National Productivity and Quality Specifications (NPQS) and** Project Specific Data (PSD) templates for preparing building project specifications through the eNPQS System (Sing & Zhong, 2001). Construction Infonet consists of articles on quarterly reviews of the Singapore construction industry, databases with details of all construction projects, construction project listings, and statistics on the construction sector—such as construction labour productivity growth, tender price index, construction costs for different types of development, and others (Building and Construction Authority, 2012).

It should be noted that the construction companies that contributed the information might not be practising intra-OL within their own companies. Moreover, the information available on these two online portals does not capture the good and bad practices of particular construction projects, nor does it cover the take-home knowledge or experience that the contractors have garnered from previous projects. Most of the inputs are just basic information or statistics, with no in-depth exploration of the projects. As there is no extant research we know of that explore OL and the Singapore construction industry, empirical research is therefore needed in this direction. Hence, the research gap is identified as the missing link between OL and productivity in the Singapore construction sector.

References

Argirys, C., & Schön, D. A. (1978). *Organizational learning: A theory of action perspective*. Reading: Addison Wesley.

Argote, L. (2012). *Organizational learning: Creating, retaining and transferring knowledge*. New York: Springer.

Argote, L., & Ophir, R. (2002). Intraorganizational learning. In J. Baum (Ed.), *Companion to organizations* (pp. 181–207). Oxford, UK: Blackwell.

Argyris, C. (1977). Double loop learning in organizations. *Harvard Business Review, 55*(5), 115–125.

Barlow, J., & Jashapara, A. (1998). Organisational learning and inter-firm "partnering" in the UK construction industry. *The Learning Organization, 5*(2), 86–98.

Bell, S. J., Whitwell, G. J., & Lukas, B. A. (2002). Schools of thought in organizational learning. *Journal of the Academy of Marketing Science, 30*(1), 70–86.

Building and Construction Authority. (2012). *Construction Productivity Roadmap*. Singapore.

Curado, C. (2006). Organisational learning and organisational design. *The Learning Organization, 13*(1), 25–48.

D'Aveni, R. A., & Macmillan, I. C. (1990). Crisis and content of managerial communications: a study of the focus of attention of top managers in surviving and failing firms. *Administrative Science Quarterly* (35), 634–657.

Dodgson, M. (1993). Organizational learning: a review of some literatures. *Organization Studies, 14*(3), 375–394.

Egbu, C. O. (2004). Managing knowledge and intellectual capital for improved organizational innovations in the construction industry: an examination of critical success factors. *Engineering, Construction and Architectural Management, 11*(5), 301–315.

Fiol, C. M., & Lyles, M. A. (1985). Organizational learning. *Academy of Management Review, 10* (4), 803–813.

Gephart, R. P., & Wolfe, R. A. (1989). Qualitative data analysis: Three microcomputer-supported approaches. In F Hoy (Ed.), *Academy of management best paper proceedings* (pp. 382–386). Academy of Management.

Guthrie, D. (2005). Organizational learning and productivity: State structure and foreign investment in the rise of the Chinese corporation. *Management and Organization Review, 1*(2), 165–195.

Hedberg, B. (1981). How organizations learn and unlearn. In P. C. Nystrom & W. H. Starbucks (Eds.), *Handbook of organizational design*. Oxford, UK: Oxford University Press.

Hong, J. (1999). Structuring for organizational learning. *The Learning Organization, 6*(4), 173–186.

Huber, G. P. (1991). Organizational learning: the contributing processes and the literatures. *Organization Science, 2*(1), 88–115.

Kamara, J. M., Augenbroe, G., Anumba, C. J., & Carrillo, P. M. (2002). Knowledge management in the architecture, engineering and construction industry. *Construction Innovation: Information, Process, Management, 2*(1), 53–67.

Kim, L. (1998). Crisis construction and organizational learning: Capability building in catching-up at Hyundai Motor. *Organization Science, 9*(4), 506–521.

Krippendorf, K. (1980). *Content analysis: An introduction to its methodology*. Beverly Hills, California: Sage.

Levitt, B., & March, J. G. (1988). Organizational learning. *Annual Review of Sociology, 14*(1), 319–340.

Morris, E. (1996). Computerized content analysis in management research: a demonstration of advantages and limitations. *Journal of Management* (20), 903–931.

Örtenblad, A. (2001). On differences between organizational learning and learning organization. *The Learning Organization, 8*(3), 125–133.

Porter, M. E. (1980). *Competitive strategy: Techniques for analyzing industries and competitors* (Republished with a new introduction, 1998). Free Press, New York.

Sing, T. F., & Zhong, Q. (2001). Construction and real estate NETwork (CORENET). *Facilities, 19*(11/12), 419–428.

Tennant, S., & Fernie, S. (2013). Organizational learning in construction supply chains. *Engineering, Construction and Architectural Management, 20*(1), 83–98.

Torbert, W. R. (1972). *Learning from experience: Toward Consciousness*. New York: Columbia University Press.

Tsang, E. W. K. (1997). Organizational learning and the learning organization: A dichotomy between descriptive and prescriptive research. *Human Relations, 50*(1), 73–89.

Tsang, E. W. (1999). Internationalization as a learning process: Singapore MNCs in China. *The Academy of Management Executive, 13(1)*, 91–101.

Annex 1: Summary of RBV Critiques

Summary and assessment of critiques to the resource-based view (RBV)

Critique	Assessment
1. The RBV has no managerial implications	Not all theories should have direct managerial implications Through its wide dissemination, the RBV has evident impact
2. The RBV implies infinite regress	Applies only to abstract mathematical theories. In an applied theory such as the RBV, levels are qualitatively different It may be fruitful to focus on the interactions between levels rather than to consider higher levels prior as a source of SCA
3. The RBV's applicability is too limited	Generalizing about uniqueness is not impossible by definition The RBV applies to small firms and startups as well, as long as they strive for an SCA Path dependency is not problematic when not taken to the extreme The RBV applies only to firms in predictable environments
4. SCA is not achievable	By including dynamic capabilities, the RBV is not purely static, though it only explains ex post, not ex ante, sources of SCA Although no CA can last forever, a focus on SCA remains useful
5. The RBV is not a theory of the firm	The RBV does not sufficiently explain why firms exist Rather than requiring it to do so, it should further develop as a theory of SCA and leave additional explanations of firm existence to TCE

(continued)

L. Sui Pheng and S. Jia Zheng, *Construction Productivity in the Multilayer Subcontracting System*, Management in the Built Environment,
https://doi.org/10.1007/978-981-13-1831-3

(continued)

Critique	Assessment
6. VRIN/O is neither necessary nor sufficient for SCA	The VRIN/O criteria are not always necessary and not always sufficient to explain a firm's SCA The RBV does not sufficiently consider the synergy within resource bundles as a source of SCA The RBV does not sufficiently recognize the role that judgment and mental models of individuals play in value assessment and creation
7. The value of a resource is too indeterminate to provide for useful theory	The current conceptualization of value turns the RBV into a trivial heuristic, an incomplete theory, or a tautology A more subjective and creative notion of value is needed
8. The definition of resource is unworkable	Definitions of resources are all inclusive The RBV does not recognize differences between resources as inputs and resources that enable the organization of such inputs There is no recognition of how different types of resources may contribute to SCA in a different manner

Note *SCA* sustained competitive advantage; *TCE* transaction cost economics; *VRIN/O* valuable, rare, inimitable, and nonsubstitutable resources and capabilities plus organization
Source Kraaijenbrink et al. (2009)

Annex 2: Interview Questionnaire for Main Contractors

Topic: The effects of subcontracting on construction productivity.

Objective: Speak to/interview some practitioners to get a first-hand feel of subcontracting practices in our industry with respect to construction productivity.

1. Does your firm sub-contract works out to others? Your answer:
2. What percentage of work does your firm typically sub-contract out to others? Your answer:
3. What types of work/trades do you typically sub-contract out to others? Your answer:
4. What guiding principles do you adopt to decide on (a) what, and (b) how much work to sub-contract out to others? Your answer:
5. In the midst of a tightening labour market and increases in foreign worker levies which has resulted in a squeeze in foreign worker numbers, do you sub-contract more works out to others than ever before? Your answer:
6. Would sub-contracting out your firm's works to others affect your project construction productivity? Your answer:
7. Do you have a core team of subcontractors whom you always work with? Why? Your answer:
8. How do you help your subcontractors to enhance their construction productivity, if any? Your answer:
9. Is your firm also a subcontractor? Your answer:
10. If your firm is also a subcontractor, what and how would you want your main contractors to help you level up your construction productivity? Your answer:

L. Sui Pheng and S. Jia Zheng, *Construction Productivity in the Multilayer Subcontracting System*, Management in the Built Environment,
https://doi.org/10.1007/978-981-13-1831-3

Annex 3: Interview Questionnaire for Subcontractors

Topic: The effects of subcontracting on construction productivity.

Objective: Speak to/interview some practitioners to get a first-hand feel of subcontracting practices in our industry with respect to construction productivity.

1. Does your firm sub-sub-contract works out to others?
Your answer:

2. What percentage of work does your firm typically sub-sub-contract out to others?
Your answer:

3. What types of work/trades do you typically sub-sub-contract out to others?
Your answer:

4. What guiding principles do you adopt to decide on (a) what, and (b) how much work to sub-sub-contract out to others?
Your answer:

5. In the midst of a tightening labour market and increases in foreign worker levies which has resulted in a squeeze in foreign worker numbers, do you sub-sub-contract more works out to others than ever before?
Your answer:

6. Would sub-sub-contracting out your firm's works to others affect your project construction productivity? How do you ensure your firm stays productive?
Your answer:

7. Do you have a core team of sub-subcontractors whom you always work with? Why?
Your answer:

8. How do you help your sub-subcontractors to enhance their construction productivity, if any?
Your answer:

9. How does your firm value-add to your main contractors whom you work with?
Your answer:

10. If your firm is a subcontractor, what and how would you want your main contractors to help you level up your construction productivity?
Your answer:

L. Sui Pheng and S. Jia Zheng, *Construction Productivity in the Multilayer Subcontracting System*, Management in the Built Environment,
https://doi.org/10.1007/978-981-13-1831-3

Annex 4: Profile of Interviewees

Interview session no.	Company role	Date	Firm type	Annex
1	Executive director	16th Jan 2016	MC	5
2	Director	18th Jan 2016	MC	6
3	Project director	20th Jan 2016	MC	7
4	Project manager	22nd Jan 2016	MC	8
5	Owner	25th Jan 2016	SC	9
6	Senior contracts manager	26th Jan 2016	MC	10
7	Managing director	28th Jan 2016	MC	11
8	Managing director	27th Jan 2016	MC	12
9	Project manager	30th Jan 2016	MC	13
10	General manager	31st Jan 2016	SC	14
11	Contracts manager	25th Jan 2016	MC	15
12	Manager	2nd Feb 2016	SC	16
13	Senior contracts manager	3rd Feb 2016	MC	17
14	Manager	4th Feb 2016	SC	18
15	Managing director	20th Feb 2016	SC	19
16	Managing director	11th Mar 2016	SC	20
17	Managing director	30th Jan 2016	SC	21
18	Managing director	12th Mar 2016	SC	22
19	Director	9th Apr 2016	SC	23
20	Managing director	15th Apr 2016	SC	24

Summary

- 20 interviews carried out; 10 Main Contractors and 10 subcontractors.
- Interview period was from 16th Jan 2016 to 15th Apr 2016.

L. Sui Pheng and S. Jia Zheng, *Construction Productivity in the Multilayer Subcontracting System*, Management in the Built Environment,
https://doi.org/10.1007/978-981-13-1831-3

Annex 5: Interview Transcript 1 (Main Contractor I1)

Date of interview: Saturday 16th Jan 2016.
Position of Interviewee: Executive Director.

1. **Does your firm sub-contract works out to others?**

Q1: Does your firm sub-contract works out to others? Like all the various trades you just mentioned?
A1: Yes.

2. **What percentage of work does your firm typically sub-contract out to others?**

Q2: What percentage of work does your firm typically sub-contract out to others?
A2: I would say in terms of labour, it is about 90% of our work. In fact, this is typical of our industry.
Q3: So most of the main contractors practised this as well?
A3: Yes, in the private sector. Unless you are saying that you are building just a factory or HDB, there is not many trades. Mainly is the structural part of the building, then maybe the structural contractor is the main contractor. So it is not necessarily the same thing. It is more of the subcontractor behaving more like the main contractor, rather than the subcontractor behaving like the subcontractor. There are many construction projects. In the industry, there are different categories. Commonly known is residential, commercial, industrial and institutions. So let's say for the residential category, private/landed, you have private/non-landed (condo) and public housing. These are the 3 main categories. The smallest segment would be private/landed. So imagine if you tender out the private/landed work to build house by the contractor, likely that the contractor will do all by himself. So in this sense, will have less sub-contractor. But when you go to the public housing segment, especially during the good old days, when you first move-in, it's the HDB unit that has no finishes, got door, window frame and the wall is painted and that's about it. In such cases, you only got the need for door, window, and tiling. So

L. Sui Pheng and S. Jia Zheng, *Construction Productivity in the Multilayer Subcontracting System*, Management in the Built Environment,
https://doi.org/10.1007/978-981-13-1831-3

mainly the scope of the work lies with the RC guy. So these RC guys, they come out from the subcontractor firms and they join the main contractor. So the situation might have improved in a way that the main contractor now is doing 70% of the work themselves and the subcontractor doing 30% of the work. But the quantum of the 30% in the HDB Project as compared to a private development 30% is different. In a private development, in terms of scope, maybe RC is 50%. Architecture and M&E remaining 50%. In the private development, the main contractor also sub-contracts out the RC work. So it's about 90% of the work subcontracted out. The remaining 10% is basically the main contractor's staff which maintains the site facilities.

Q4: So that part of the 10% is from the main contractor?

A4: Yes that part of the 10%.

Q5: So their role is to manage the sub-contractors?

A5: It is more to provide the backup like cleaning, safety, general maintenance.

Q6: So does the people in this office now counted within the 10%?

A6: The people in this office don't count. They are not counted into the foreign worker numbers. The people in this office are supervisory staff. If you ask me on the supervisory staff, our supervisory staff, compared to the subcontractor's supervisory staff, for every one they have, we also have 1 or maybe we have 2. If you add up all the sub-contracting staff, it may reach 50 or 60. For our side, it can be 80 or 90 staff. These are all supervisory or management level. But for workers, for every 1000 workers, 900 belong to the subcontractor. For foreign workers, what I feel is that. So that's this issue of overlapping. It doesn't mean that once we get the contractor, we subcontract out 90% of our work and then we give others an impression that we did only 10% of the work. We still do 100% of the management. The work we sub it out, but the speciality of the main contractor is management. Even though that worker is the subcontractor's worker and even though it is the sub-contractor's routine, we are able to control it as though it is our routine.

3. **What types of work/trades do you typically sub-contract out to others?**

Q7: What types of work/trades do you typically sub-contract out to others? E.g. Like M&E etc.

A7: Don't constraint your definition of specialist works to just M&E. Any Architecture trade or RC trade is also a specialist. (It could be anything other than the management of the entire project site, progress etc.)

4. **Is your firm also a subcontractor?**

Q8: Is your firm also a subcontractor?

A8: No, we are a main contractor.

5. **What guiding principles do you adopt to decide on (a) what, and (b) how much work to sub-contract out to others?**

Q9: What guiding principles do you adopt to decide on (a) what, and (b) how much work to sub-contract out to others?

A9: It would definitely be what type of work which would be carried out. So let's say for example, company A which is a main contractor, it is constantly only participating in HDB market. Hence, it only tenders for HDB projects exclusively. So HDB project 1, HDB Project 2 and 3 are similar. Since company A is determined to have such a fixed requirement, then it makes sense for them to develop their own team of workers so that they cut out one middle man. This is because if you hire one more subcontractor, you need to pay for the subcontractor's boss and got overhead. Hence you save on the boss and overhead as the worker cost is the same. However, there are some companies like ours on the other extreme, one job to the next job is totally different. So now, there is no continuity. Hence in such situation, it will make no sense to actually maintain a particular team of worker with specific skillset. For each of the different jobs, we need different skillsets. So that's why, for our type of business, for every job, depending on the skillset necessary, we will identify and match the correct specialist subcontractor and engage that subcontractor.

Q10: So the main key focus is still on the skillset which you are looking for that particular project?

A10: That skillset is required to do the job that the company is tendering for; that type of job that the company does. Hence, there are the 2 extremes that I have given you. One is contractor that only does HDB projects while the other contractor does not constraint itself to a particular market segment. They are participating in multiple market segment. As such, these firms need to be agile.

Q11: Hence, I can say that your company is flexible and agile?

A11: We are extremely flexible and we are really at the extreme.

Q12: So your company can literally tender for all jobs?

A12: Yes, we can tender for all jobs. Well, there is pros and cons but over the years, we have turned out that way. Somehow, we have managed to survive in this sort of situation.

Q13: Can I deduced that other main contractors of similar size such as your competitors they also have the same guiding principles, to match the skillsets of their project and thus they know how much to subcontract. Is this usually the case in our industry?

A13: Yes, this is because if you always do a particular job and you always do this particular work on your site, then it may be worthwhile to invest for the longer term. Same for equipment, certain equipment if you use it job after job, you will just buy the equipment.

6. **In the midst of a tightening labour market and increases in foreign worker levies which has resulted in a squeeze in foreign worker numbers, do you sub-contract more works out to others than ever before?**

Q14: Recently in Singapore, the Singapore Government actually shrink the number of labour permits given to foreign nationals such as the S-Passes. This has resulted in a tightening labour market. Therefore, do you subcontract your work more than before because of this circumstance?

A14: It's still the same as before.

Q15: So even though there is a shrinkage in the pool of foreign labourers available, it doesn't affect how much work you sub-contract?

A15: No. This is because we don't have a choice. Firstly, there is only foreign worker, there is no such thing as Singapore worker. So you can shrink whatever quota, the worker will still be 100% foreigner. Hence, if you shrink the pool, there is no advantage of getting your own worker. On the other hand, which is also related to your earlier question, because company like ours or any other companies, unless you have a continuity of work that requires the same skillset, then it will result in redundancy when you don't have another project that requires workers of the same skillset. On the other hand, the subcontractor have many avenues as they have many clients. So that is the efficiency using the subcontractor. In fact, even with the MYE reduction, it will be more efficient to use the subcontractors as there are able to shuffle around to get work for themselves.

7. **Would sub-contracting out your firm's works to others affect your project construction productivity?**

Q16: When you subcontract your work out, does it affect your construction productivity?

A16: I would say for our case, it will be better productivity. This is because the subcontractors are also known as specialist as they are more skilled.

Q17: Hence, what is your definition of project construction productivity?

A17: The work that has to be continuous. Meaning that the labour being applied, can be applied continuously and the work pace can be further ramped up. This is because if you deploy a team of workers, then they come in and start working, no stoppages, then you know they can build up momentum. Hence, they can achieve high productivity. If they come in and then there is lots of issues, changes such as site obstructions, then you cannot build momentum. Therefore resulting in low productivity.

Q18: Hence your idea of construction productivity is the need for continuity and momentum.

A18: Yes.

Q19: As for the other main contractors, do they also feel the same way as well?

A19: I think so. At the end of the day, it all boils down to this dominating KPI called "Making money" (Revenue generating) You can say that if you are productive, you can make more money. If you are not productive, you cannot be making money… or rather you cannot make money when you are less productive. So just imagine that a worker just come in to work in Singapore and the worker can only work that fast. For example, if this worker worked 10 units per day, who is working close to the maximum and then you apply some technology that increase that work to 12 units per day. However, the worker can only work half a day because the other half a day he is waiting for something. So you only get 6 units. Hence, there is this problem of stoppages and continuity. The other issue is technology where technology can increase the units from 10 to 15 units per day. If that is possible, that technology will definitely be used. This is because it will enable the company to make more money. However in reality, there is only that few technologies that are available. There is a few technology to make formwork faster and other technologies to do rebar faster. Even so, there is a lot of other work that is on-site that is not available despite the existing technologies that are available in the industry. Hence, there is limited benefit or potential using the technology. Definitely, when there is technology that can be used, it will be used.

Q20: So in that sense, most of the main contractors in a bid to ensure higher productivity, would rather subcontract than to employ technology?

A20: No, I would say that contractor firms essentially still operate like a business. There will always choose the best due to market forces. If by subcontracting they can make more money, then definitely they will subcontract. If they do it themselves, they will incur higher risks and mistakes will be made easier. Then there might be stoppages that will affect work progress. If hire subcontractors and there is a stoppage, main contractors need not pay the subcontractors at all. The subcontractors will pay for their own workers if that occurs. Hence for main contractors, if there is a stoppage, then you would still have to pay for your own workers if the main contractors did not subcontract. Hence, it really is not so much on the issue of procurement, the other aspect you can consider for your research is… maybe one argument against too much procurement is that of the training and capability. This is because if you have your own workers, you can invest in their training. Whereas you have subcontractor's workers, you may not be able to insist on them going for training. Again, that will depend on the critical question of what kind of training are you talking about? Is there really some training that is so good that will enable your workers to be "superman"? If it is really that obvious, then yes workers will have to go for that training. Again, you will have to come back to the same question of do I still need that worker's skillset for long? If I do not require their skillset for long, I cannot afford to keep them for long.

Q21: Therefore, it is much more cost effective to subcontract out the works?

A21: Yes.

Q22: What is your take on the BCA construction productivity indicator? Is it bothering a lot of contractors?

A22: I think that BCA's principle/philosophy of productivity is to try to do more with less with less resources or more mechanisation. Or rather, with more mechanisation, to replace the traditional methods. Hence, using less manpower. Therefore, BCA's principle is really compressed into that thought. For me, there is another concept; that is to do less in the first place, by cutting wastage. That is also a means to productivity which is much cheaper, but not say much "cheaper", but maybe we haven't be using resources fully. I think it is common that our industry don't care about wastage. So wastage can come in many forms. If you do more than it's necessary, so obviously you are not going to be productive right? Even if you can do work efficiently, but you are doing more than its necessary. This results in a loss of productivity, even though you do it very efficiently. You redo work when it is damage or when it needs to be changed. For example, it is quite common that the designer change their minds, hence you need to redo that particular work again. Or if you are doing on a task, then some parties want to come and inspect; which is very common in construction because 20 years ago, we went through an evolution of quality control where it is carried out in the 1990s. First don't talk about productivity, talk about quality, where ISO 9000 is the standard for quality. Hence, lots of progress works are due to stop for inspections etc. in the 1990s at institution projects. As such, there are multiple inspections resulting in many stoppages. If there are 20 workers, all of them have to stop work and wait for 1 supervisor to inspect. As such, if 20 workers wait for one hour, then it would be 20 man-days already, hence it is time-wastage. So there are many such kind of waste going on. So this waste, if you eliminate it, I think that amount of productivity that you save, easily will match the productivity gained through mechanisation and it really costs less, greener and time saving. Without doing such time-saving or by not addressing wastage, it will also hinder the benefits that brings along with mechanisation, even if mechanisation is employed at its optimum. This is because machine can also be stopped due to the need to wait, start-up. In addition, regulation can also cause wastage.

Q23: Therefore, at the end of the day, the contractors in Singapore are less inclined to adopt BCA's way of mechanisation but rather they will focus more on cutting down wastage which helps to improve productivity?

A23: Not really because whatever mechanisation BCA is encouraging is useful. It is about looking at the whole picture. The technology available is only applicable to certain part of the project. The rest of the project there is no new technology available. Hence, just by applying those few technologies is not the magic bullet. It doesn't mean that so long as you apply mechanisation, you are going to achieve the productivity target. So what more do we need to do? My point is that to do more, we have to do less. If you do less, that means you will end up being more productive. It's just like doing ISO 14001, now they talk about SMART technology and systems. These are all active measures. On the other hand, when a room is designed to be well-ventilated, say lots of shading, then that is passive. Hence, it is cheaper and more environmentally friendly. To achieve this, you don't need to do anything like installing a SMART system etc! Don't need to put anything there in the first place. So you plan out everything properly, more smooth and efficient inspection and don't make any changes, you will be more productive. Hence, this part on wastage, or the Japanese called it Kaizen or lean, where this aspect has not been taken into account seriously. I think this area is a huge untapped area. I think we are starting to get in to it. Hence, to do more, you got to get less. This is because in the industry, there are a lot of things going on. Whatever we have now the mechanisation, can only impact a fraction of productivity. I'm not saying it's bad to adopt technology, it's good technology but it doesn't apply throughout the whole spectrum. You cannot use the tool that the RC (Structural) guy is using and to employ to another types of work like tiling etc.

Q24: So that means BCA must redefine or expand their scope on productivity?

A24: I think the productivity measurements is still the same. We can make everybody more aware of where we have waste. And this industry is one that has lots of waste. If you ask me why the construction industry is lagging behind other industry, why our construction industry lags behind other countries' industry is due to such wastage. If you study industry production and construction productivity, they are all about not only efficiency but also on cutting down wastage. They adopt machines to be faster but what are they actually talking about? You learnt in your course about all the problems associated with mechanisation right? Queuing, jams up in the downstream process, also known as bottleneck. So these management issues can affect productivity as well. Now, BCA focus on the mechanisation but as seen in our assembly line, some parts can do by machine, the remaining other parts still have to be done manually. Do note that we are not factory production that produces only one product, we in construction industry is much more complicated than that.

Q25: As such, we have to look at construction productivity at its entirety and not restrict ourselves to just adopting mechanisation to improve construction productivity?

A25: Yes. It's a long process and journey for us to actually strive to improve our construction productivity.

8. **Do you have a core team of subcontractors whom you always work with? Why?**

Q26: Do you have a core team of subcontractors whom you always work with? Why?

A26: We do have a group of subcontractors but we are also open to engaging new subcontractors. Pros of core subs- familiarity and predictability. Cons-tend to have lock in i.e. Subcontractors expect higher price for same job. Another issue here is complacency, where new subcontractors can also provide new skills to the setup and keep existing subcontractors on their toes.

9. **How do you, as main contractors, help your subcontractors level up to enhance productivity, if any?**

Q27: How do you help your subcontractors level up to enhance productivity, if any?

A27: We keep them abreast of advances in methods materials and equipment. Help them identify areas for improvement and training opportunities.

10. **And if you are subcontractors, what and how would you want your main contractors to help you level up for construction productivity?**

Not applicable.

End of Verbatim Report 1

Annex 6: Interview Transcript 2 (Main Contractor I2)

Date of interview: Monday 18th Jan 2016.
Position of Interviewee: Director.

1. **Does your firm sub-contract works out to others?**

   A1: Does my firm sub-contract works out to others, the answer is yes.
   Q1: Ok.

2. **What percentage of work does your firm typically sub-contract out to others?**

   Q2: What percentage of work does your firm typically sub-contract out?

   A2: Well, I normally do condominium works. Hence, you got more ivy work like kitchens, wardrobes and appliances. So the percentage could be a bit higher. So the figure could be somewhat like 35% subcontract out to the nominated subcontractors (NSCs) like for mechanical, electrical, wardrobe, windows etc. Another 35% subcontract out to domestic subcontractors. Hence, it is a good 70%. Preliminaries in a contract usually ranges 15% in the contract. Then another 10–15% is structural works like formwork, rebar, concreting, drainage which we normally do ourselves. These will give you a good figure of the contract sum. Hence, the various contract value allocation altogether adds up to 100%. So actually, the main contractor's job is more on project management. Ours is really project management.

   Q3: So it is pretty common for most main contractors to sub-out up to 90% of their work?

   A3: I would say about maximum 75–80%, because preliminaries constitute a good 10–15% of the project already. Another point is that most contractors will do work like concreting themselves. Hence, I would say a good 75% will be sub-contracted out. This is because 15–18% are normally the preliminaries. Preliminaries are for the project manager's job to consider insurance, overheads, drainage, supervisor's safety, salaries, site safety etc.

L. Sui Pheng and S. Jia Zheng, *Construction Productivity in the Multilayer Subcontracting System*, Management in the Built Environment,
https://doi.org/10.1007/978-981-13-1831-3

Q4: So all these are the main contractor's job. Am I right? They seldom sub-contract all these out I suppose?

A4: Yup, they seldom do so. Even if they do, then it equates to total sub-out and then they collect 3–5%.

Q5: Oh I see. Ok.

3. **What types of work/trades do you typically sub-contract out to others?**

Q6: What types of trade do you usually sub-contract to your sub-cons?

A6: I will give you a range of things that I normally sub-contract out. For the NSCs, I will give you the list of works that we sub-contract to them. For the domestic subcontractors, we subcontract the works such as doors, plastering, painting, brickwork, tiling, ceiling boards, screeding, timber floors and sanitary.

Q7: So the domestic subcontractors are usually hired by the main contractors themselves?

A7: Yes, that's correct.

4. **Is your firm also a subcontractor?**

Q8: Are you a subcontractor?

A8: We are not a subcontractor.

5. **What guiding principles do you adopt to decide on (a) what, and (b) how much work to sub-contract out to others?**

Q9: What are your guiding principles which you adopt to decide on (a) what, and (b) how much work to sub-contract out?

A9: For those works that we do not have expertise in like door frame, tiling, we don't do it in-house. Also, when we decide to subcontract out tiling works, we will subcontract all tiling works out to the subcontractors. Supply of the tiles is by us, the main contractor. The subcontractors will provide the labour to install these tiles.

Q10: Could it be due to the difficulty of securing foreign labourers in the midst of the tightening labour market which prompted you to subcontract out?

A10: It is not because of the shrinking labour pool, but the subcontractor is one that has worked with you (main contractors) for a long time. You don't choose a subcontract because its new. We also look at the track record as well. Once I realise I don't have the expertise for a particular work, I don't try to get workers to do it as it will involve my employees who need to know the tradecraft which is not that simple. Having the knowledge and skill is different. I may have the knowledge of how to do a good tiling work, but that doesn't mean I am skilled at laying tiles. Therefore, knowledge is not equals to skill and this goes vice versa.

Q11: This means that you would rather go for a subcontractor that has the hands-on experience to complete that particular trade?

A11: Of course, that's the only way. The guiding principle for us is that the subcontractors must have a track record, we know them long enough (Prior working experience and relationship) and by word-of mouth recommendation. We will also interview these subcontractors; see their previous job sites etc. Sometimes, we look at their paid-up capital of the subcontractor.

Q12: What do you mean by paid-up capital of the subcontractor's firm?

A12: It is the worth of the company, which means how much they invest in the company. Some subcontracting firms have only $2 paid up capital while others have $2 million. Chances are for firms that have $2 paid-up capital, it tends to be a fly-by-night company and if they run into any problems, they will just let it go. So such requisite criteria ensure that the subcontractor firm is financially sound.

6. **In the midst of a tightening labour market and increases in foreign worker levies which has resulted in a squeeze in foreign worker numbers, do you sub-contract more works out to others than ever before?**

Q13: With the ongoing tightening labour market which resulted in a reduction of foreign worker numbers, does your firm sub-contract more works out than before?

A13: No, it's still the same. Even though many contractors are complaining about this circumstance.

7. **Would sub-contracting out your firm's works to others affect your project construction productivity?**

Q14: Would sub-contracting out the works to others affect your project construction productivity?

A14: The answer is yes and no. Yes in the sense that if the subcontractor is no good in doing the trade, then it will slow down productivity.

Q15: Meaning that the subcontractor is unable to handle the job?

A15: I think you should rephrase your topic. This is because one of the things is that you failed to realise that subcontractors are lower tier contractors, so their management system is not as sophisticated as the main contractor. When you talk about the effects of subcontracting, this is in fact the guiding principle. It is not all these. Let me try and rephrase this question of yours. The main contractors are professional and more literate, you know more management principles, tend to attend more seminars, courses to understand and learn. Coming from the main contractor, some of the main contractor's companies are actually managed by subcontractor firms that have become main contractors. This is because when they grow big, their capital increases, then they total subcontract from the main contractor to do a job of value e.g. $20 million. Then they recognised $10 million. Hence, they upgrade that way and later become main contractors. Some main contractors actually adopted this route. So when comparing between main contractors against the subcontractors, I would say that those subcontractors with lesser educational qualifications are unable to grasp the impact of not going for courses, seminars etc. In fact, it is harder to convince them to go for such courses as they see themselves as doing financially very well-off thus far, hence lacked the motivation and drive to upgrade themselves. However, these subcontractors cannot understand that there is a better method as they cannot grasp. The constant search for a better value-added solution but sadly they cannot understand. Such methods are Just In-Time solutions (JIT) which can help to improve productivity. Sadly, these contractors do not understand the value and purpose of JIT and that by not planning, they can bring total chaos. If we can do a time-study cycle, you will ask yourself why do we need so much workers to do this part of the work? The deployment of workers is very important and the question is you send the workers every day for tiling. Then have firms wondered how many m^2 that they are supposed to produce? If you don't know how many m^2 supposed to produce, how do you know what is the number of workers to be deployed? If you don't know, then how do you know what type of works has been sub-contract from and what ought to be paid does not tally already. If you sub-contract for 1 m^2 for $10, then your workers, your levy can cost up to $200 per day for example. Then in this case, you have to produce 20 m^2 right? Do note that you have yet to make money as you have not paid for office overhead, supervisor staff, your boss salary. So in other words, when you deploy your worker, do you know how much your worker should produce? If you don't know, then it's going to be a problem right? To these subcontractors, their nonchalant attitude with regards to how many workers to be deployed for a particular trade, where they blindly guess the number of workers required for a particular job. Hence, this results in inefficiency, also known as wastage. Go read the book written by Professor Lim Lan Yuan on wastages. Wastages is anything that does not contribute to productivity. It can be waste in motion, where the worker forgets to bring his tools to his job site and needs to retrieve it again elsewhere. It must be advocated that every skill worker must carry their toolbox, like that of the Aussies. Our workforce here has a lot of problems and when you want to study the effect of subcontracting, it is not subcontracting per se, it is the mind-set of doing things. Wastages is very important. The moment you can control wastage, surely productivity will increase. These are the different types of waste in motion; abortive work, not doing right the first time as doing right the first time has many implications. Hence, there are 3 things you need to know—1. Levels 2. Dimensions 3. verticality. These are the 3 cardinal principles of construction.

Take an example, for pile caps, whatever it is, you need to do excavation correctly. Hence, when you cast concrete, you get your desired pile cap. As such, you need to get your levels right from the start. Next example, the dimensions for your bathroom must be right. If it's wrong, then you will not be able to fit your bathtub in. Lastly, verticality is paramount in structural works such as walls, columns and lift shafts. Example would be if it is not vertical, the guide rails of the lift cannot be installed. Hence, rework needs to be carried out by hacking the lift walls, resulting in wastages. Other implications such as hacking and trimming that needs to be done on the surrounding structural members like slabs. Hence, it is paramount to get it right the first time in terms of the dimensions and verticality. Of course, the finishing product must be good as well. So if you look at effects of subcontracting, it has nothing to do with the labour shortage issues that we commonly heard about that plagues our construction productivity.

Q16: So it's more on the practices of the industry?

A16: Yes. Like the mind-set etc. Firstly, when you do something, you must set out to plan. Besides planning, in order to succeed, there must be scenario painting. This is an army term used to plan what is going to come out. Hence, you must be able to envisage, anticipate what could go wrong. Therefore, before what could go wrong, you can check before any work commences. As such, if any error is spotted, you can rectify immediately or risk big huge chain effect. So the envisaging and the anticipating phase, you are drawing knowledge from your past experience. It's like your alert book as to what may go wrong using your past 30 years of experience. So the question is to sit down and discuss and talking, Understand?

Q17: So basically it is to run through with various parties, in terms of coordination, am I right?

A17: Who are the working people do you need for your job, as you must understand that most of them are formerly farmers who have been planting wheat and paddy for 20 years of their life and then you ask them to become a skilled construction worker. What is our understanding of a good quality work? India and Singapore is of different quality, simply because we are a first-world country. To them, water can flow is good enough already, right? Expectations are different. So therefore, when a worker is asked to set out to work, the same worker you must know and observe how he does his work. You don't simply leave your worker to work by themselves right?

Q18: Obviously not. You need to supervise them.

A18: Well, it happens and its very common. Having told a worker and having to supervise the worker closely is different. That's why LKY has 3 rules. Firstly, nothing is impossible. Secondly, you do not assume the worker will do it—Don't Assume. Thirdly, you must not say you do not know, you must go and find out. We are trying to draw from his second philosophy—Do not assume that the workers will do it. That's his philosophy, do not assume! So actually construction is mere engineering. It is a lot of psychological. You agree?

Q19: So it's all about the perception of people, their mind set. Am I right?

A19: How do you teach people the trade? How do you motivate people? You understand what I'm saying?

Q20: Ok. So it's more on dealing with people.

A20: You are trying to gauge a person's ability to see where he will go wrong. And then that will avoid abortive work. Understand? So your focus is a bit off.

Q21: The focus of my research is a bit off?

A21: You should not say "effects of subcontracting" but when you say effects, you can break it up into 2 parts. One is low productivity and the other is knowledge of the subcontractors. The professionalism of subcontractors, understand?

Q22: Ok. Noted.

8. **Do you have a core team of subcontractors whom you always work with? Why?**

Q23: Do you have a core team of subcontractors that you usually work with? Why is this so?

A23: Yes, definitely. This is because of the quality, expectation, management expertise and it is really a working relationship already. Also because of credibility and reliability as well.

Q24: So you only go for trusted subcontractors and in the event that these subcontractors also have their hands full but you still have more jobs that you want to subcontract, do you go out and source for new subcontractors?

A24: Of course. But that seldom happens. We try not to do that. Easily, my subcontractor will come back and look for us.

9. **How do you help your subcontractors level up to enhance productivity, if any?**

Q25: How do you help your subcontractors to level up to enhance productivity?

A25: This is a very tough question. This is always a constant process of monitoring, talking to them, understanding them. I think at the end of the day, a lot depends on our management. It all depends on discipline in their work, discipline in studying details carefully like we as main contractors must get our levels and dimensions correct, give clear instructions to our subcontractor and follow up with that by checking on what they have done. Also constant interactions with them.

Q26: Other than the various modes of communication which you have adopted such as email, whatsapp, phone calls etc., do you adopt technology such as BIM etc.?

A26: BIM is very important. Definitely. But how many subcontractors understand BIM? Very little right? A lot of main contractors also don't understand BIM.

Q27: Is it because all still do things the old-fashioned way.

A27: Yes.

Q28: What do you understand of project construction productivity?

A28: Firstly, productivity is inversely proportional to wastage. That defines productivity. Once wastage goes down, productivity goes higher. Firstly, must bring down wastages. The biggest culprit is wastages. You understand? So what are wastages? Rework, waste in motion, not getting clear instructions, no directions, workers and machines idle, these are all waste. Not getting ready tools and equipment on time. For example, excavation of trench on the road. If the lorry don't come, how to excavate? So all the workers wait then! You can think of any wastages and the key to handling productivity is wastages!

Q29: Recently, BCA is driving the industry towards mechanisation and off-site production…

A29: That I agree though. However, there is a point of diminishing returns. Can everything be off-site?

Q30: Not everything though.

A30: A lot of people like standardisation of toilets like that of HDB. But when you buy a $5-million-dollar condominium, can the toilet be the same as HDB? Obviously not! The value will depreciate. Will all the toilet be the same? Some people like marble, some like granite. So how? At the end of the day, we must accept that certain things can be off-site. The biggest one is toilet where you have the most trade; ceiling board, all these and that, a lot of things. Toilet can be off-site.

Q31: What about mechanisation?

A31: Of course can. But depends on how big your area is. Factory is possible where you will be casting 1000 m^2. On your HDB flat, how are you going to use robots? Hence, there are some limitations. I will still think that certain things can do it while others can't. I will still think that to do all these is to bring down wastages. If we can define what is wastages, then we wouldn't have a problem right?

Q32: So is BCA's definition of productivity alright?

A32: I wouldn't say that it is narrow. I would start off by saying that the contractor must be knowledgeable, must understand work cycle, time study of items and must have the tools and knowledge.

Q33: Does this imply new methods of working?

A33: I wouldn't say new methods of working as the method is always there but you refine the methodology. You refine the time cycle. For example, when do you need 10 workers? Can we get 10 workers?

Q34: Meaning that we have to conduct forecasting of demand etc.

A34: Yes, you must get ready your things isn't it?

10. **And if you are a subcontractor, what and how would you want your main contractors to help you level up for construction productivity?**

Not applicable.

Additional Comments by interviewee

Q35: Oh ok.

A35: Actually, all the focus on subcontracting practices is basically wrong approach already.

Q36: Wrong direction you mean?

A36: Wrong direction! That's why if you ask me, my presentation very psychological. Nothing to do with equipment and nothing to do with robots. Therefore, trust, clear instruction and direction. Trust—To be treated equally with the domestic subcontractors. So it's best to manage by working. Hence, using WhatsApp tool to talk about attendance, all about mutual trust and respect. So basically, must get everybody on the same boat. With that, things get done easier.

Q37: So I can say that the root cause of low productivity that is plaguing the industry is due to poor attitude of the industry professionals itself. Not being able to embrace change etc.

A37: Yes.

Q38: So like you mentioned, once you tackle the psychological part, the technical part becomes easier?

A38: Yes. So a recap, do you know the difference between skill and knowledge?

Q39: Yes I think. So skill is the hands-on experience that is required to deliver the work. Whereas knowledge is an idea of how to do something.

A39: Just give you an idea on painting. Some people have the knowledge, but unable to do painting. So skill is the application of the knowledge and knowledge is what you gained via studying and research.

Q40: So skill is a higher tier than knowledge?

A40: Yes, definitely. Such concept is very hard for people to grasp.

Q41: Why not complement skill and knowledge together since it can work well together?

A41: You can. I have the knowledge where I would know how to do it. My workers are the skill.

Q42: Hence at the end of the day, it is still about the collaboration and cooperation needed by all parties.

A42: Correct. I think I am the first contract to tell you about attitude and it is fundamental to this industry. I am in the construction industry for so many years, and I think all these problems that are related to the industry is all due to the psychological aspect. It's all about the mind.

Q43: Oh alright.

End of Verbatim Report 2

Annex 7: Interview Transcript 3 (Main Contractor I3)

Date of interview: Wednesday 20th Jan 2016.
Position of Interviewee: Project Director.

1. **Does your firm sub-contract works out to others?**

Q1: Does your firm actually sub-contract works out?

A1: In all construction, yes. Let me give you an example. Main contractor usually got two trades that they like to do themselves. 1. Structural and 2. Wet Trade.

Wet Trade there are many types. Plastering, brickwork, blockwork etc. For structural and wet trades, some of the companies love to do themselves. The rest of the other trades they will sub-out. Because structural and wet trades, they can control the progress of the works. Those that they sub-out like dry trade, M&E.

Q2: So what are the types of dry-trade?

A2: Dry-wall, ceiling panels. Why they want to sub out this? Because it's all about the money. Material can be costlier. For wet trade, it's got cement involved so the cement content they know the price very well. They know the cost of water, cement, which are controlled prices. Hence, they know. For dry trade, should the Architect want to pick a particular material and it cost $5, then want to change to another material that cost $20. Hence, we don't specialise in this as we are the main contractor. Similarly, structural works are controlled prices; rebar, formwork are all controlled prices. In the industrial practice, they usually love to do this. However, got some companies also sub-out total—100% as a main contractor. Some main contractor will like to do structural and wet trade by ourselves. For example, for us we do the structural trade in-house. The wet trade we sub-out.

Q3: Oh ok. So just now you mention the wet trade consist of work like plastering, formwork, tiling etc. while dry trades are like those of dry-wall etc.

A3: Yes. Structural work consists of 3 types; Formwork, rebar, concrete. Wet Trade will involve tiling, hence we sub-out as there are many different types of tiles and we don't know the price.

L. Sui Pheng and S. Jia Zheng, *Construction Productivity in the Multilayer Subcontracting System*, Management in the Built Environment,
https://doi.org/10.1007/978-981-13-1831-3

Q4: So we sub-out so as to lower the cost and also lack of expertise in doing this, hence easier to manage.

A4: Yes.

2. **What percentage of work does your firm typically sub-contract out to others?**

Q5: What percentage of work does your firm sub-out to subcontractors?

A5: There are so many trades, 50 of them and we only do 1—Structural works. Maybe only 10% we do ourselves.

Q6: Only do 10% of the work in-house, so remaining 90% sub-contract out?

A6: Yes. Because nobody wants to do the remaining 90% as the risk might be high. As the job gets bigger, some main contractor doesn't like to take the risk. For me, I will sub everything out. Structural can also sub-out. Structural work can consist of a lot of things; sub-structure, superstructure etc. Substructure can mean ERSS—Earth Retaining Structure System. These are the expertise we don't have. These are all substructures. Basement work, CBP wall, piling work all sub-structure. All these constitute 10% only.

Q7: So the remaining 90% all are sub-out?

A7: Yes.

Q8: Is such a practice pretty common in our industry?

A8: Yes, definitely. Because of labour cost also. The labour cost is also a killer… you are paying a $550–$750 per month per labourer back to the government. So why are you taking that type of risk?

Q9: So to minimise the risk and cost, you sub-out almost everything.

A9: Yes, that's the whole idea.

3. **What types of work/trades do you typically sub-contract out to others?**

Q10: So the types of trades you typically sub-contract out to others are what you mentioned earlier, am I right?

A10: Yup. So for us it's just wet trade, M&E and sub-structure work.

4. **Is your firm also a subcontractor?**

Q11: So is your firm also a subcontractor?

A11: Well, that depends.

Q12: Is it because your firm started out as a sub-contractor then slowly upgrade to become a main contractor?

A12: Yes. At A1 financial standing.

Q13: But how does a firm actually slowly upgrade from subcontractor to become a main contractor?

A13: We have to do a private job. Some companies love to do private jobs rather than do government jobs. In order to upgrade themselves to do government jobs, they have to do certain percentage of private jobs to upgrade their rating and financial pay-out. Then that's when from a C1 financial standing become B then A.

Q14: So approximately what's the time taken to upgrade the firm from C financial standing to A standing?

A14: 5–7 years easily. Some takes longer, depending on the economy outlook as Singapore is very small. Limited space as well hence limited projects.

Q15: Earlier in the interview, why did you imply that as a main contractor, you can be considered as a sub-contractor? This is because your initial response to Q4 is that it depends…

A15: It depends on the whole contract package on how the government awards it. For example like mega projects such as MBS, RWS, there so many components.

Q16: So it is because of the sheer magnitude of the project, then even though you are a main contractor, you are considered a subcontractor in that instance.

A16: Yes, it depends on how you look at it. Like RWS, we end up as part of the main holding of contractors.

Q17: So it's a matter of perspective?

A17: Yes, it is.

5. **What guiding principles do you adopt to decide on (a) what, and (b) how much work to sub-contract out to others?**

Q18: What are your guiding principles to determine what to subcontract and how much work to subcontract?

A18: Usually we look at the whole entire building work, also known as (aka) total management.

Q19: So you rather have control over the entire building than letting other people do?

A19: Yes, it's total management.

Q20: Which means if you can handle, you handle the work by yourself. If you can't, you subcontract out. Am I right?

A20: Yes, well you see the guiding principle here is that for example: If company A has a subsidiary company, you call it Company B where Company A has certain percentage of shares in it. When company A gets the contract, then it can give it to B to do. Of course, between company A and B, there must be a contract binding between them. Then A sub-out to B, then B does the job on A's behalf.

Q21: So it's more of a commercial viability point of view? To ensure that the subsidiary stays commercial healthy?

A21: Yes, as both A and B are related. Because A might have 10 projects, then if one more comes along, then it gets B to do for them. Then this is considered a total management system undertaken by B. It can be everything… From M&E, to entire structure and external work. So its total management done by B.

Q22: Hence, your firm practices total management? And what is the name of your subsidiary firm?

A22: Yes, we practice total management and our main contractor is Ken-PAL. They got a subsidiary called PAL-Link. So I'm actually doing this total management and a lot of companies are actually doing it to actually reduce the risk. It is also to build up both parties' standing and we also got the experience in performing all these tasks.

Q23: Besides this total management system, any other guiding principles?

A23: No. Usually it's the risk factor, minimising risk.

Q24: Hence, I would say that your key guiding principle is to minimise risk at the end of the day, am I right?

A24: Yup. Nowadays like for the lift upgrading programme, a lot of main contractor after they got the job, they allow the subcontractor to come in to use their name and do the HDB job. Say for example, I'm A and you are B. You come and approach me to ask whether I can give you the job. Then I say ok then you help me bid for the job. Once you got it, you are using A's name, my name, to do the job.

Q25: So how do they make such an arrangement possible?

A25: There must be some form of profit-sharing between me and you. Whatever you have done the work, the paymaster (Owner) will pay me and then I pay you. This is because the job is bid under my name. So both parties must have the trust and to understand each other. So for example when you bid, you got the job of $20 million, of course you won't get the full $20 million. You are using my name. Hence, for example for the 1st month for the job duration of 20 months, you claim $1 million, the money will come to me first. When I pay you, it is not $1 million but you will receive $850,000.

Q26: So it's some form of profit sharing.

A26: Yes. I won't see you die and you won't see me die. Then this is how you are going to total management it.

Q27: So many of the main contractors and subcontractors they have such an understanding?

A27: Yes.

Q28: Will it resort to the subcontractors merging with the main contractors eventually?

A28: Sometimes it does, then if I'm A (Main contractor) then I will ask you B (subcontractor) to do this job for me if I have more jobs since you are doing so well.

Q29: So many subcontractors get absorb into the main contractor?

A29: Yes, so the job that I pass to you, I pay you separately at different rates as this is clearly my job. I am just making use of you as a subcontractor. Whereas as mention earlier, that one you are using my name to bid, it's a different thing. So if this is the case, it is very healthy for both you and me. So the relationship and trust grows, if I cannot find anybody, I can find you and let you do the job.

Q30: So if that's the case, the subcontractor is known as domestic subcontractor?

A30: You can call it like that but we usually call it total management system. When you work with me, it's call total management system. Anybody engage under your own contract is then called domestic subcontractor. When you tie contract to contract, I don't call it domestic subcontractor. It's called total management contract. That means the manpower comes from you, even the project manager comes from you and you are the boss. Between A and B, you don't call it domestic. Anything you engage I don't bother, it is called domestic. Anything between A and B, we call it total management system or contract. Why? Because all the PM, staff belongs to you and not me. Anything that comes under you, it's called domestic sub-con.

6. **In the midst of a tightening labour market and increases in foreign worker levies which has resulted in a squeeze in foreign worker numbers, do you sub-contract more works out to others than ever before?**

Q31: Now under the midst of the tightening labour market like the reduction of S-Passes etc., do you sub-contract more works out even than before?

A31: No, it's still the same. But then you have to find very innovative ways of doing things even though you are a subcontractor.

Q32: Can you explain more on this?

A32: Make use of more machinery. Some of these jobs can just make use of more machinery, make sure the subcontractors reduce the waste and debris. You want to speed up construction, use machinery. You want to speed up certain process, like structural, use formwork system; less cutting. So these are some of the things that as the main contractor will have to innovate. You just have to innovate. I give you an example: If the designer call for CBP (Continuous Bored Pile) walls. The diameter is roughly around 800 mm. Then they have another diameter of CBP of 300 mm. But then to do it faster, you can take out the 300 mm ones. Maybe don't even need to have the 300 mm CBP. Must know how to innovate. We use CBP here for AS8 but we innovate it. The back row we reduce it. So a lot of things like formwork, we innovate, due to the labour shortage. So like formwork systems, to reduce the striking time, precast, our link bridge from RC work we change to composite panels to reduce my formwork. Cos Formwork, tying rebar and casting needs manpower. So if I use composite, up goes the I-beam and I lay the deck, thus I don't need the formwork.

Q33: Won't construction cost go up if we use such innovative methods and even be riskier in terms of the financial aspect?

A33: Of course, some of them will. There will always be risk, these are calculated risks.

Q34: So long as you can deal with the risk then it's fine, correct?

A34: Yup.

7. **Would sub-contracting out your firm's works to others affect your project construction productivity?**

Q35: Would sub-contracting your firm's work affect your construction productivity?

A35: No. You just have to track daily the work procedure vs. the work schedule. The number of work today that needs to be done you put in manpower, based on your schedule. If you complete, then that's it.

Q36: So just make sure you follow according what you have planned beforehand?

A36: Yes. You need to more or less really know how to innovate and track. In fact, right now I'm doing an assignment from BCA on the construction and productivity project.

Q37: So regarding construction productivity, is there any challenge to improving it?

A37: Relooking into the design. Sometimes it's too massive and need to simplify. We are trying to look at the whole entire project… An easier way of doing construction. Look at the design first then the manpower. Certain things like change of formwork, then change your manpower. Look at the system first where there are lots of parts to look into. Sometimes by saving certain time, it doesn't mean that you reduce labour to save certain time. Time itself you can save with the same amount of worker, then it's good enough. An example where the job needs 6 men to be done and you have to do it within 6 days. But if you can have 6 men but finish the job within 4 days, that itself is an increase in productivity. Look at the process, not at the human factor.

Q38: So ultimately it can be both ways, the time and human factors as well in tackling productivity?

A38: Yes. Say the process itself takes 6 people, don't try to reduce manpower. You can push your subcontractor to shorten certain process.

Q39: Hence, at the end of the day, even though you sub-out the work, it doesn't affect you're your project construction productivity because of the active monitoring and control of your subcontractor and with ample communication with their supervisor to ensure they have what they need to complete the job?

A39: Yes. That's right. Very important.

Q40: Anything else to take note of to improve project construction productivity for main contractors?

A40: When you want to talk about productivity, always look at the process, procedure because it is always done with manpower, material and machinery. I tell you one secret formula that I always practice: M^2, S^2, T^2.

Look at the material, then the manpower. This is followed by the step and then with the sequence. This is followed by the correct tool used and the timing.

Q41: Ah! So that's the key right?

A41: Yes! That's the key word. Because all these, manpower and material are cost. Step and sequence are not cost. It is the process, again I emphasize! Step and sequence is the process, then with the tools and timing. If you get this right, you shorten every step and sequence, you save already! Instead of doing things for 6 days, you do it in 4 days. Save already! This is the secret formula that I have been using it.

Q42: I interviewed other main contractors as well and they say that to improve productivity, we have to cut down on wastage. Is that right? So wastage is a huge problem in our industry?

A42: That's correct! Formwork, tiling and plywood you cut and throw away, a form of wastage.

Q43: Do you see any method as of now to reduce wastage?

A43: Hmm… At the present moment for Singapore, it is a pitiful thing.

Q44: Is it because no one is practising it?

A44: I wouldn't say no one but as I say… we are relying on foreigners. Different culture and mentality comes into it. A project manager or even a staff, they can only do so much to achieve so much. And to contribute to the waste, it all boils down to the design factor. From the architect and consultant. Similar to safety, why are there so much accidents even with so much emphasis from the government? Again because there is a cultural difference between foreign and local practice, hence accidents will still occur. Certain things cannot be contained, the more roads you build, you don't expect that there will be no road accidents. There will still be. It has nothing to do with the number of roads built. It is the present system that has a failure to it. There is no such thing as zero accidents. Of course like wastages, a lot of people wanted to control. They will be glad if say a $10 million projcct, wc can control the wastages to $100,000. You are then very good already and it is very difficult already. Because those things that we buy, use and then throw away, we cannot reuse. In fact, we cannot reuse. For example the road space, there is a fixed dimension of 1.8 m width. Then if we need to factor in the joints because the Architect wants and his design, then we have no choice but to cut and throw away if the overall width exceeds 1.8 m for the tiles. No choice. You buy, then you throw away even though you have paid the supplier for the tiles already because the tiles are cut already. It is really a waste!

8. **Do they have a core team of subcontractors whom they always work with? Why?**

Q45: Do you have a core team of subcontractors whom you always employ and why do you keep rehiring them?

A45: Yes. The wet trade sub-con has been with us for 3–4 years. The structural sub-cons have been with us for 7–8 years already. It is a matter of trust as you become worried if you get in another sub-con as to whether can they perform or cannot perform.

Q46: Initially, when you first subcontract out, how do you choose which sub-con can best complete your job?

A46: Again, it's a matter of trust. Certain project we have to try out other sub-con to see how good they are.

9. **How do the main contractors help their subcontractors level up to enhance productivity, if any?**

Q47: How do you help your subcontractors to level up to enhance productivity, if any?

A47: For what we have here, we actually change the formwork system and for some of the work we counter-propose to lessen down the sub-con work. Usually, changes are from the main contractor. For sub-con, we only push for the productivity from the sub-con with all these changes. So the sub-con knows that there are certain changes, say with lesser workers being used.

Q48: So it means that you as the main contractor, you vet through the sub-con's work plan?

A48: Yes, definitely. We have to be one step ahead. A sub-con cannot do that; they just propose how to work only. Here at NUS AS8, we adapt the system formwork and we adapt the composite panel to speed up the work. It is not even proposed by the owner. We had to fight for it.

Q49: So if you are happy with the plan, you allow them to continue. If not, you make changes to it. Even if it is higher cost but save on time? Am I right?

A49: Yes.

Q50: So do you keep your subcontractor inform of the latest developments in trends and technology?

A50: No, we don't. Because sub-cons they come and go.

10. **And if they are subcontractors, what and how would they want their main contractors to help them level up for construction productivity?**

A51: How would they want their main contractors to help them level up construction productivity? I am very sure the subcontractors cannot achieve this.

Q52: Is it because of the mind-set, attitude and educational qualification?

A52: Yes, correct. The government is acting across the board, whether you are main contractor or sub-con. If you are the sub-contractor, you make sure you train your guy. If you are the main-contractor, you still have to train your guy and must even train more. If you are sub-con and doing specialist trade, you still have to train them. There is a fund for you to improve your trade called PIP. You look at the whole entire list then you know which trade you are coming from, for both the main-cons and sub-cons. We don't push the sub-con for it because the sub-con knows about it at this moment of time. Whether they choose to get a machine to split the work from BCA or we choose the machine to work and pass it to you. The sub-con will never change the system form. We will change the system form. So hopefully, from then, with the number of manpower, the same manpower, we can split up the work, rather than face with a situation where there is not enough workers and need to bring in more workers. We will discuss with the sub-con to limit the number of workers, where there is adequate manpower to do the job. If the sub-con still requests for more manpower for a particular work, then we will change the work process such that the job can be done with the present number of workers without bringing in additional workers.

Q53: So the number of workers remain, just that the method of working changes.

A53: Yes, so that's why I keep saying the process. The worker in one day can only do so much. You can multi-task but you can only do so much. Between 8 and 10 h, you can only do so much.

Q54: So basically it's all about the process?

A54: It all about the process. You look at your process. If you are spending time talking to me, it's part of the process. Whether is it healthy for you or more for today, we don't know. It's part of the process. In Singapore, in order to really increase productivity, if you want, go back to design and build. Contractor hates direct contact with the owner. From here (D&B), we appoint architect, structural engineer, appoint M&E and then contractor we have our own QS department.

Q55: So design and build is one that can save time and cost?

A55: Yes, as we as the contractor then know what the owner wants, not the architect that tries to manage everything.

Q56: So the traditional design-bid-build is inefficient?

A56: Yes. Hence, design and build is the way. This is because the traditional design-bid-build got the clerk-of-work and resident technical officer, all doesn't work! They are just here to check paper work as to whether you got submit or not.

Q57: They check on whose behalf? The owner, am I right?

A57: Yes. The traditional way is owner, example NUS gets the architect to do the design. From the design, they also appoint a structural to look at the design. Then they also get the M&E to look at the design.

Q58: So then you liaise with multiple parties and that slows down the entire process?

A58: All talk about design first. All design finish, then they tender. One main-con have to report to architect, M&E… etc.

Q59: So the current arrangement with NUS AS8 project is also the traditional way?

A59: Yes. They got the RTO, RE to assist all these 3 parties (Architect, Structural and M&E), to look at the main contractor. Like this how to increase productivity?

Q60: So in a way, it's like too many hierarchy?

A60: Yes, that is the worst thing. If the contractor understands what the owner want, we build according to what the owner wants. We tell and bring the architect along and tell him that since he is under our payroll now, please address what the owner wants and we build together. Can you imagine the traditional way, how tedious it is.

Q61: Besides being tedious, it can easily lead to miscommunication?

A61: Yes. Then under the owner, still got NSC, then they also have their direct sub-con. So can you see how many trades they have here?

Q62: Multiple trades then.

Additional Phase

A63: Well there is another way you can write for your paper. It is about the role and responsibility of an RTO. What does the role of RTO play? What do they do in terms of: 1. at the site, and 2. in the contract?

Q64: Don't the RTO report to the owner?

A64: Well that's the issue. I don't know. It's so grey! Are they here to oversee the project to be completed on time? Are they here to act for the owner or the consultant. So grey! They just dump the RTO in just because the contract calls for.

Q65: Is it the legislation that calls for RTO to be included in such projects?

A65: It's the registration that has to be done as regulated by BCA. These RTO needs to be registered. What is the role of RTO? These people are redundant! If the Architect and the main contractor go and see the owner, then we know what the owner wants, then why do you need to engage the RTO to check on our design if we go hand in hand? Why do you need to engage another guy? It's stupid, am I right? That means the owner has no faith in you. Waste money and cause a drop in productivity. So a lot of time this occurs, so you should write about this and a lot of time, the locals don't like to become RE. When students graduate, why graduates don't like to become RE? Why? Only the foreigners like to become RE.

Q66: Is it because of job uncertainty?

A66: Well, I would like you to address this part. A very good topic to write. The role and responsibility of RTO and RE in the construction industry. Do they form part of the sub-contracting of the project? If it exists, which part is it? Are they representing owner or consultant? Are they also helping the owner and consultant to improve productivity besides the main contractor?

Q67: Alright. In addition, is it more productive to actually retain trained workers so that you spend less time to retrain new workers?

A67: Yes. If I'm not mistaken, a few years back the work permit for workers is 2 years. Easily the training for a worker for 2 years work permit is close to $1200–$1500 for 12 months. Then, some of the workers after 2 years they go back home, then result in waste of money. So now the permit is 1 year, the training for the worker now ranges from $600 to $800, because some of the safety supervisor course and first-aid is over a few hundred dollars. Now the permit is shortened to 12 months, we still have allocated aside a budget for career training for the workers, staff such as safety and risk assessment training, BIM training. Thus we have training budget for new guys and old guys. We still have to train you even if you have to leave after 3, 4 years. It is a norm. We cannot find locals but only foreigners.

End of Verbatim Report 3

Annex 8: Interview Transcript 4 (Main Contractor I4)

Date of interview: Friday 22nd Jan 2016.
Position of Interviewee: Project Manager.

1. **Does your firm sub-contract works out to others?**

Q1: Does your firm actually sub-contract works out to the subcontractors?
A1: Yes, we do.

2. **What percentage of work does your firm typically sub-contract out to others?**

Q2: So what is the percentage of works your firm typically sub-contract out?
A2: Normally about 20–40% and it depends on the contract amount, scope of work, time period given and the lowest pricing offered by the sub-contractor. It varies from project to project. Every project is different and hence, it depends on the type of contract work that we got. Sometimes we also 100% subcontract out. For example, JTC dictates that only Class A1 contractors can tender for a particular job and then there are only 2 such firms in the market that bid this job. However, since we got the job and then the job is all about dredging, but we do not have the expertise in dredging. Hence, we subcontract fully out to a specialised subcontractor who has the expertise to do dredging. Nonetheless, we will monitor them closely to ensure that they still complete the job on time for handover. For this case, we usually subcontract out to our core team of subcontractors that are deemed as more reliable. Therefore, at the end of the day, it all depends on the contract.

3. **What types of work/trades do you typically sub-contract out to others?**

Q3: So what are the types of trades that you typically sub-contract out?
A3: We subcontract out these types of trades, as follows:

- RC and precast work
- Dredging work
- Piping works

L. Sui Pheng and S. Jia Zheng, *Construction Productivity in the Multilayer Subcontracting System*, Management in the Built Environment,
https://doi.org/10.1007/978-981-13-1831-3

- Steel pipes and spun concrete pile
- Pile-testing (Normally do Statnamic load test and PDA* Test)
- Instrumentation testing
- Diver works (underwater welding etc.)

(Note: *denotes Pile Driving Analyser).

4. **What guiding principles do you adopt to decide on (a) what, and (b) how much work to sub-contract out to others?**

Q4: What are your guiding principles to determine what to subcontract and how much work to subcontract?

A4: Our guiding principles to decide what to subcontract and how much to subcontract are as follows:

- Specialisation of work
- Scope of work highly technical and difficult to complete (Hard Layer dredging)
- Quality control, in the sense that we choose contractors that we know of and whether we have the in-house capabilities to deliver the job
- Allocated project time. If project duration too short, we subcontract out
- Workload of the firm at that point in time. If too much to do in-house, we also subcontract out
- Most importantly, the chosen subcontractor must have enough relevant experience
- Pricing, usually go for the cheapest. We call it lowest complying bid.

5. **In the midst of a tightening labour market and increases in foreign worker levies which has resulted in a squeeze in foreign worker numbers, do you sub-contract more works out to others than ever before?**

Q5: Now under the midst of the tightening labour market like the reduction of foreign labour quota etc., do you sub-contract even more works out than before?

A5: Still the same as before. However, having said that, we subcontract out because we want to reduce risk by subcontracting only to contractors with experience and they are indeed specialised in their trade. In addition, we also keep track of our budget so that it does not exceed as allocated before.

6. **Would sub-contracting out your firm's works to others affect your project construction productivity?**

Q6: Would sub-contracting your firm's work affect your construction productivity?

A6: Yes. It will improve construction productivity as these subcontractors are more familiar in their field of work. They have the capacity to complete that particular job. In addition, to ensure efficiency, best to enable easy communication between you and the subcontractors. This means that we have to coordinate and manage effectively.

7. **Do they have a core team of subcontractors whom they always work with? Why?**

Q7: Do you have a core team of subcontractors whom you always employ and why do you keep rehiring them?

A7: Yes. The subcontractors are usually from the trades that is mentioned in question 3. We choose them because they have the expertise, they have good management of their work, good working relationship, sound company finances and well-experienced as they have done many of such jobs before.

Q8: In the event that all your 6 core subcontractors are also busy, do you still continue to look for other new subcontractors?

A8: Yes, we do.

Q9: How do you evaluate the new subcontractors which you intend to sub-out to?

A9: By using the same evaluation criteria like our 6 core subcontractors. It is mainly the lowest price, enough experience and have the expertise to do so.

8. **How do the main contractors help their subcontractors level up to enhance productivity, if any?**

Q10: How do you help your 6 subcontractors to level up to enhance productivity, if any?

A10: It's good to have good coordination and proper management with the subcontractors, updates and daily meetings to convey the right information. Sometimes, it also comes down to how understanding the client can be, as it cannot be just the main contractors that are pushing for construction productivity. The clients/owners must also be receptive of it. Some of the measures you see just now during your site visit such as our on-site prefabrication yard, it has greatly helped us in achieving productivity, despite the need for trailers to transport the prefab units for a short distance from dock J10–J11. Initially, our client don't even give us space for this prefabrication yard as they deemed it to be unnecessary and they are not supportive of our productivity efforts. For us, we have to fight for it, convince their management to accept our proposed method of relocating the prefabrication yard on-site that will greatly improve productivity. It took them a while to give the go-ahead. Last time, another client as well, they are also not receptive of our methods to improve productivity, work methods that can save time. For this other client of ours, they are more interested in making money and ensuring their operations are not disrupted at all. Their management scared that they will incur losses even with our modified proposed method of working more efficiently with minimal disruptions. So yes, it can be very tiring and draggy when the clients themselves don't get the buy-in and the rationale of our work method, even if it is enabling us to work more efficiently and thus improving productivity.

A11: In order to improve productivity, government must keep skilful worker by issuing longer S-Pass, especially those in more specialised job. This is to minimise resource wastage and financial loss if the worker who underwent training and worked in Singapore for a year that must go back home, just because the S-Pass is almost expiring already. Waste time and money as well. In addition, widen the job scope and its availability in order to keep these skilled foreign labourers. Lastly, more funding from the government can go a long way to help lower the firm's cost to ensure financial stability for the firm.

9. **Is your firm also a subcontractor?**

Q12: Is your firm also a subcontractor?
A12: Yes, we can be considered as a subcontractor. We are normally main contractor, about 95% main contractor, 5% subcontractor. We specialise in marine constructions like jetty, harbour, berth and bridges. Occasionally, we become subcontractor to do heavy lifting such as barge lifting and marine piling using drills such as the reverse circulation drill etc.

10. **And if they are subcontractors, what and how would they want their main contractors to help them level up for construction productivity?**

Q13: So with regards to the subcontractor, what and how do you think the subcontractors want their main contractors to help them level up for construction productivity?
A13: On the part of the main contractors, we will meet with the subcontractors regularly to close the gap. Besides that, we will also recommend our subcontractors to go for relevant courses as required to upgrade their knowledge and skills.
End of Verbatim Report 4

Annex 9: Interview Transcript 5 (Subcontractor I5)

Date of interview: Monday 25th Jan 2016.
Position of interviewee: Owner (Subcontractor specialising in Duct Fabrication and kitchen and exhaust cleaning).

1. **Does your firm sub-contract works out to others?**

Q1: Does your firm, a subcontracting firm, sub-contract works out to other subcontractors?

A1: Yes, I do and I subcontract everything out as I have no more core competency. Because now the labour cost is so expensive, I gave away all my workers.

Q2: What do you mean by "gave away all your workers"?

A2: I started in kitchen exhaust cleaning and I have a few of such jobs. Then the levy kept going up and I couldn't get enough jobs but I have already build up my track record already with people coming to me asking me for work. Then I don't have enough work, thus I gave all my workers to another kitchen exhaust cleaning company. Thus, I actually subcontract out everything and now I am just a marketing agent. In conclusion, I sub-out everything as I have no more core competency already.

Q3: Can you tell me more about your firm and what it does?

A3: My firm started as a kitchen exhaust cleaning and we had a bunch of workers and we also do a bit of duct fabrication. Then we had not enough customers as we took a long time to build up our customer base. So it's not easy starting from scratch. A lot of firms in this line they are already in the industry and then when they set up, they already had customers and then they get the workers. So I did backwards, I got the workers first, so I ended up paying them month after month up to a year. So that's about it. Then after almost 2 years, I decided to give up the workers, "renting" out the workers to other firms to survive and then I just couldn't do it anymore. So in the end, I gave out the 4 workers that I had to another kitchen exhaust cleaning company or Fabrication Company. I don't do the work anymore. So when customer comes, I just sub it out to another firm.

L. Sui Pheng and S. Jia Zheng, *Construction Productivity in the Multilayer Subcontracting System*, Management in the Built Environment,
https://doi.org/10.1007/978-981-13-1831-3

2. **What percentage of work does your firm typically sub-contract out to others?**

Q4: So even though your firm still exist but you will sub it out totally to another contractor of the same field?

A4: Yes. That's right, 100% subcontract out. Then I will tell that firm that I am here to just make money so that firm have to give me a rate that I can make money. So I sub everything out. So it will screw up your data. In the midst of the tightening labour market, I have no more workers already.

3. **What types of work/trades do you typically sub-contract out to others?**

Q5: So what are the types of trades that you typically sub-contract out?

A5: I only subcontract out jobs such as duct fabrication and kitchen exhaust cleaning.

4. **What guiding principles do you adopt to decide on (a) what, and (b) how much work to sub-contract out to others?**

Q6: So for your case, any guiding principles to determine what to subcontract and how much work to subcontract?

A6: Again, I am here to just make money. The firm have to give me a rate that I can make money and I sub everything out. Next is lack of manpower, where a lot of firms are squeezed out from the market and they gave up their workers when they are unable to extend their work permits. Hence, this results in the sub-contract practice where subcontracting firms that do not have enough labourers will sub-contract to those that have adequate labourers.

5. **In the midst of a tightening labour market and increases in foreign worker levies which has resulted in a squeeze in foreign worker numbers, do you sub-contract more works out to others than ever before?**

Q7: Now under the midst of the tightening labour market like the reduction of foreign labour quota etc., do you sub-contract even more works out than before as a subcontractor?

A7: Yes, you have to sub even more. I have spoken to the firm that I gave out my workers to, all the works from that firm have been subcontracted out. You can't subcontract to him. Now a lot of firms are squeezed out from the market and they gave up their workers. So this firm that I refer to, he runs an 8–12 men team, basically just got subcontract works to do every day. He got very few of his direct customers, he just "eat" other people's subcontract only…

Q8: So the business naturally comes to him?

A8: Yes. Well, the market is consolidated, where everybody just drop-off their workers and so the only 1 guy that has got the workers will get the job.

Q9: So it means that whoever has the skilled labourer wins?

A9: Yes, correct. So the firm with the labourers quote me a job, then I quote the customer and then I make maybe $300–$500. Something like that. Sometimes if you are worried, then you can get your men that you have just gave-up your workers to another firm to wear your firm's T-shirt and then go and do the job.

6. **Would sub-contracting out your firm's works to others affect your project construction productivity?**

Q10: Would sub-contracting your firm's work, such as your duct fabrication, to another subcontractor firm, would it affect your construction productivity?

A10: Actually, it doesn't. I find it much better to subcontract out the work. So that you don't have the liability of handling the dormitory, workers' fight, pay issues etc. Last time very troublesome, have to handle all these. Such as go MOM, go medical check-up etc. Now you sub the work out, you just think of how much you make, how much it cost. So it's actually quite productive. When I receive a customer's call, I will send "my men" down. Well actually, it is the boss of the firm that you subcontract your job to. So nowadays, I don't even show up at the customer's place, just stick to this arrangement. So for example, the boss of the firm actually goes down to the customer's place, then he quotes me $600. Then in turn, I will call back the customer and quote the customer $900. So it's a very efficient process. I only do paper work now where I just generate invoice and quotations. Hence, it is efficient. Last time my workers have to go down to the site to see for themselves, report to me etc. Now, I trust this subcontracting firm of mine so much that I do not worry that they will "run away" or steal my business from me. The boss of this firm depends on all of these "marketing agents" to market for him. So the market is quite funny. The one that can do the work is very few but the ones that advertise that they can do the work are many firms.

So it is the ironic situation of this labour crunch which is very stupid. Ideally, everybody wants their own labourers as it is cheaper, but that's ideal case. However, things in reality are not always ideal. I give you an example… this water proofing company called Proof-tech. Proof-tech is only about 5 years old but they manage to grow into a size so ridiculously big, like 20–30 lorries now, approximately 200–300 workers now just solely doing on waterproofing and they advertise very aggressively. Few companies that do waterproofing in Singapore like ES etc. There're 4 or 5 of them and they all go back to the same office. So some of these companies are GST registered, some not GST registered, depending on say when they all bid for a condominium job. These companies, although different, are all under the same boss. Very aggressive. When you open the yellow pages, the joke is that when you look at the phone numbers, it is all the same boss even though different phone numbers. So waterproofing is already cornered by PT. I think they are the 2nd largest in Singapore and now, when you open up "The Contractor's Directory", when you look at this book, go look at waterproofing works. You look at the address of these different firms and they are quite close. Maybe it's like next door units! So when I approach the boss of these firms and say that I don't want to pay GST, the boss said ok and he used another company that is not GST registered and I got the invoice free of GST. So you see it's very powerful. Another trade is plumbing where I know of this boss that got 7 hand phones which corresponds to 7 different plumbing companies. If you Google on Singapore's plumbing firms, the top 6 is his.

Q11: Wow! So they are dominating the market?

A11: Yes! That plumbing boss I just mentioned, he's got 30 plumbers running around Singapore.

7. **Do they have a core team of subcontractors whom they always work with? Why?**

Q12: Do you have a core team of subcontractors whom you always employ?

A12: Yes. I have to use. Usually, I make a deal with the subcontractor where I give the subcon this job and this subcon must ensure that they make money constantly as I will give all the jobs that I have to them. All of plumbing job, ceiling panel job, painting job, I always have a specialise guy for each. The only problem was the subcon get the extra competency like last time do painting then now can do timber decking… Then I have a headache as when they increase their core competency, I am caught between this sub-con and my preferred sub-con for timber decking jobs. So these are the worries that I have. Typically, I make a deal with them to pay them on a per job basis at present. In the past, I would "squeeze people" by asking these sub-cons to give me a bulk rate as I guarantee them "x" amount of jobs. In failing to provide the required "x" amount of jobs to these subcontractors, I top up to them. So I prefer that the subcons tell me how much and when I make the margin, it's done deal.

Q13: So in this sense, lesser work and hassle for you right?

A13: Yes, so in that sense, it is efficient but you make a lot less.

8. **How do the main contractors help their subcontractors level up to enhance productivity, if any?**

Q14: How do you help your subcontractors whom you always subcontract works to, assist them to level up in order to enhance productivity, if any?

A14: No, I don't do that.

Q15: So this has never crossed your mind?

A15: No. You don't really care what their skill level is, so long as they can do the job will do. Because a lot of times, hitting paperwork is not a requirement. In fact, the key is that a plumber need not need to be a licensed plumber to work. In fact, a licenced plumber need not know how to do plumbing! They only do stamping! In the good old days, licenced plumber very easy to get. Now, the licenced plumber only end up do endorsements. They look at the work of this plumber, scan through his documents and then do stamping (endorsements) of the documents. It's a joke man! This is the same for LEW (Licenced Electrical Workers). The only difference is that LEW knows how to do work, unlike the licenced plumber! I give you an example, plumber will quote me $250. I would quote the customer $500. I would tell the plumber to go down and check. Then the plumber gave feedback to me regarding the plumbing problems. When the customer asks me how much, then I quote $500. I simply make the difference. Especially for emergency cases such as I/C choke, I can make even more money!

9. **Is your firm also a subcontractor?**

(Not applicable, firm is a subcontractor.)

10. **If your firm is also a subcontractor, what and how would you want your main contractors to help you level up your construction productivity?**

Q16: So as a subcontractor, what and how do you want your main contractors to help you level up for construction productivity?

A16: The biggest problem for main contractor and subcontractor is payment. A lot of times the firm that I subcontract my works to, they are in trouble now as his firm tried to borrow money from me. He did a duct fabrication job for a condominium. The duct job price is about $60,000. Usually, you must ask for a down payment. The main contractor agreed to only give $10,000 as down payment. Then this subcontractor ask for half of the duct job price, $30,000. Then the main contractor said that if they did not accept, they will give the job to someone else. So as a subcontractor, he scared of losing business. Hence, they did it. That's all the payment he ever got. So he finished the job at a cost price of $55,000 and his margin is only 20%. So one month of work he get $15,000, still not too bad. However, all he got was $10,000. So now he's got problems. What happened to be main contractor was the main con was late on delivery, hence got LD (Liquidated Damages). The main con paid the LD but did not pay the subcon. So therefore, some subcontractors like Proof-tech do not want to work under main contractor. If you want their services, they must be NSC (Nominated Sub-Contractor). Hence, they deal directly with the main con. They cannot be the domestic sub-con because of such payment issues. So for the duct job subcontractor, what happened was the main contractor get into LD, then tried to slam it at this sub-con. However, the reason for LD is other works that are not done on time, not the fault of duct-works. Nonetheless, the main contractor insisted that if the main con gets into LD, all parties will LD together!

Therefore, if you get a main-con that is mean, then it will definitely affect the sub-contractor's productivity. Now, this poor sub-contractor is now chasing for the main con for payment every day. Now, he's wasting a lot of time trying to get his payment. Then he's spending a lot of time chasing for payment when he should be going out there trying to get new business!

In a nutshell, the main contractors have to pay the subcontractors on time. Usually, most sub-cons deliver work that is good. Rarely do they give shoddy work. I think the key is paying on time. There was once I did a job for PUB and it was concreting. It was civil engineering work and I didn't know how to do it. So I hired a few Chinese nationals from Geylang to help me do it for me. There was formwork etc. Then I gave PUB the payment schedule and tell them that if they don't have 40% down payment to me, I won't start the job. They paid. Subsequently, for the interim payments, they are quite punctual. You have no choice but to this as you have already engaged the workers and you have already accept the price and 'my' workers are not my workers actually, they are sub-con workers. I block book them for 18 days already. I cannot make unnecessary changes as they are with me for these 18 days and they must finish this job. So each worker that I hired was $2000++, on a rental basis. So it's the cash flow issues that subcontractors faced.

As a subcontractor, I do not want to be bogged down by the thinking that I would not be paid. This is because my productivity gets affected as I have got to keep thinking when would I get paid. This is because as sub-cons, I would need to pay my suppliers, so this definitely affects productivity. If sub-con gets paid on time, he's happy. Even though that sub-con didn't make money, it doesn't matter to him as he's got paid on time. The key thing is that if you spend time trying to recover money, then that is not your core competency already. Your core competency as a subcontractor is what specialised trades you are proficient in tackling, not collecting money. That's my take.

Q17: So it is not so much on the various methodology adopted in pushing for higher productivity but more on the fundamentals, such as the main contractors fulfilling their obligation to pay the subcontractors on time for their work done properly without any delay?

A17: It's like the main contractors can give them a payment timeline, where if they subcontractors cannot finish work by a specific date on the timeline, then they will get LD. Hence, this will spur the sub-con to work very hard, to be more productive, such as pulling manpower from other sites to do your job. If still cannot, I will pull manpower from other sources, even from your competitors! All so for the sake of not incurring LD. Payment is very important. Usually when a main contractor subcontracts out a job to a subcontractor, they don't want to think about anything. The main contractor set aside a sum of money for the subcontractor to finish the job and they don't want the sub-con to give them headache and they must finish the job. Main contractor usually talks like that.

Sub-con will think of a way to make a profit and how to think of ways to finish the job within a given timeframe. They will also when to think of the processes in order to cut down on manpower yet think of suitable methodologies that can complete the job within the shortest possible time or at most, within the given timeframe. That's what the sub-con will be thinking. The sub-con must also source for suitable materials at the lowest price that fulfils the specifications given and they have to start bargaining and whether is there a deal with the suppliers. Thus, the job of a sub-con is to deliver on-time within the given budget and to make money from it. Then the main contractor doesn't really need to think, just move on to the next work that needs to be subcontracted. Then, the main-con will tell their Project Managers to manage and coordinate the works carried by the sub-cons and ensure no clashes in works.

Hence, the role of the project managers is to ensure that the different trades don't clash with one another. It is important to conduct effective scheduling and sequencing to ensure that the works that should be carried out are not disrupted or interrupted unnecessary to ensure smooth completion of the entire project. On a small scale, there will be bickering if such work clashes occurred between the various trades. On a bigger scale at larger sites, it can easily result to fights, surely you wouldn't want all these to happen, especially when there are huge machineries within the vicinity. Hence, tradesman cannot clash with each other. If got clash, we die!

Additional Points

A18: From the perspective of the main contractors, they don't really want to care much. The main con has so much problems handling the clients when they request for multiple changes. At the same time, the main contractor must relay the instructions passed down from the client's consultant on certain works that needs to be done in the building. Then the main con must liaise with its sub-con and determine if their sub-con can do the job and negotiate for an acceptable contract value agreeable by both main-con and sub-con. If both parties (main con and sub-con) are set on the contract value, they will enter into a contract and commence works. Note that Project Managers (PM) don't know the prices. It is the QS (Quantity Surveyor) that knows the price. PM only knows who do what, when are the sub-cons coming etc. PM is the guy that does the coordination and that's the realm of the PM. Main con don't like to bother with what the sub-con is doing but the main-con has a project manager, RTO (Resident Technical Officer), RE (Resident Engineer), etc. So the RTO is hired by the main contractor, whose job is to make sure that the works are done properly. Besides the RTO, RE is under the client where their job is to check on the main contractor and ensure works are done properly in accordance to specifications and standards.

RE's job is to check on the main-con. There are a few RTOs below the RE; namely structural RTO, M&E RTO, P&S RTO etc. The main contractor also got their own RTOS. Mind you that they are not friends as the main contractor's side is to make sure that the works are done properly and he signs off. Then he brings the work to the client's RTO. Then once all works in order after client's RTO check, then it will be sent to client's RE for final checks. Once client's RE signs off, then the job will keep moving. RE's job is just to represent the client. On our side (Client's side) is that whenever we have a meeting on site, we will ask the RE to report to us and brief us on the progress of the work schedule. I once told the main-con to de-rust the rebars. The RE has the power to do such things.

But to be honest, for productivity, from that perspective, productivity can also increase if the PM is very experienced. How so? For example, for concrete casting works on a bigger scale, the PM is an experienced individual, he will roughly know how much time plumbing works is going to take, then he can arrange for the next set of trade to come in and do the work. So you can see that an experienced PM can tell how much time is required for each trade and their respective sequence required. All these are foreplaned accurately and are communicated to the respective subcontractors. So this will affect productivity on site. The Project Manager from the Main contractor side has this big role as he don't want his company to get LD. The PM can easily screw it up by arranging the works such that they have too much lag time, then this is bad. The best way to arrange and plan is for jobs to big in consecutive order. No lag time. This will also improve productivity.

Usually for large jobs, there are 2 Project Managers. One from each side. The client side PM will set the direction and the stages of work that ought to be completed within a specific timeframe. The main contractor's PM is the crucial guy that plans how to achieve and what types of work that needs to be done within the specific timeframe and making sure that the main contractor doesn't get LD.

Q19: Do you view the increased number of subcontracting works as a hindrance to productivity?

A19: Well, I tell you what's the problem of having your own workers. For example, you have your own workers doing a particular trade. Then that particular job is not available. So what do you do? If the core competency of your workers is to do duct fabrication and the job is not available to them, then in order to maximise your workers, you ask them to do duct cleaning as well. This actually happened to my core subcontractor (sub-subcontractor). Even though this is not the core strength of the workers, but he rather do this as he don't want to give his workers idling. Keep them occupied and reduce his cost. However, this is not productive as it is not using the correct worker for the correct job.

Q20: And this is quite common in the industry?

A20: Yes, definitely. It's simply the right man for the right job that also contributes to construction productivity.

End of Verbatim Report 5

Annex 10: Interview Transcript 6 (Main Contractor I6)

Date of interview: Tuesday 26th Jan 2016.
Position of interviewee: Senior Contracts Manager.

1. **Does your firm sub-contract works out to others?**

Q1: Does your firm actually sub-contract works out to the subcontractors?
A1: Yes.

2. **What percentage of work does your firm typically sub-contract out to others?**

Q2: So what is the percentage of works your firm typically sub-contract out?
A2: We try to subcontract out as much as possible. Our ideal target is 100%.

3. **What types of work/trades do you typically sub-contract out to others?**

Q3: So what are the types of trades that you typically sub-contract out?
A3: We usually subcontract out all the trades if possible, especially so when we managed to sub-out 100% to our subcontractors. From there, we will closely manage these subcontractors to make sure they get their respective jobs done.

4. **What guiding principles do you adopt to decide on (a) what, and (b) how much work to sub-contract out to others?**

Q4: What are your guiding principles to determine what to subcontract and how much work to subcontract?
A4: Do note that we are just a general main contractor. As such, we rely heavily on subcontracting to specialist subcontractors. Our guiding principles for subcontracting are as follows:

- Subcontractors must specialise in a particular trade. They are also known as specialist subcontractors.
- Subcontractors must have the capacity in terms of manpower and resources, as well as not bogged down by many other jobs and can commit to our projects.

L. Sui Pheng and S. Jia Zheng, *Construction Productivity in the Multilayer Subcontracting System*, Management in the Built Environment,
https://doi.org/10.1007/978-981-13-1831-3

- They are financially sound and not run into cash-flow problems.
- They have a good track record for that particular trade they specialised in.
- They can produce work of good quality by adopting quality control practices.
- Must meet regulatory requirements, such as having BizSafe certification etc. However, it is preferred that they are of BCA A1 financial standing.

5. **In the midst of a tightening labour market and increases in foreign worker levies which has resulted in a squeeze in foreign worker numbers, do you sub-contract more works out to others than ever before?**

Q5: Now under the midst of the tightening labour market like the reduction of foreign labour quota etc., do you sub-contract even more works out than before?

A5: Still the same as before. Even in such difficult times, work still has to be done right? Job must still be finish on time. We keep very few general workers for themselves. We usually distribute the MYEs amongst all our subcontractors to assist them in getting enough skilled labourers for them to complete the job on time. With the MYEs allocated to our subcontractors, they will then come to our site and do the work. Despite the reduction in the MYE (Man-Year Entitlements) for all contractors, we still must complete the job within the stipulated timeframe. In the case where particular trades such as concreting and formwork has too large a scope for a single subcontractor to handle, we will spilt these trades out into a few portions and ask the chosen subcontractors to do finish the job. This is to ensure that the subcontractor has ample capacity to get the job done on time without straining them. In addition, we pay our subcontractors according to the amount of work done. No main-con will pay their sub-con by man-hour. If we pay by man-hour like one of the main contractor in the industry did, the main contractor sure go bust like what has happened to a particular firm. Their subcontractor actually tried to slow down the works as they are paid by man-hour. This ended up with delays in the project and resulted in high amounts of LD incurred by the main-con. Eventually, that main-con cannot financially sustain itself and it went bankrupt.

6. **Would sub-contracting out your firm's works to others affect your project construction productivity?**

Q6: Would sub-contracting your firm's work affect your construction productivity?

A6: Yes. It will improve construction productivity as these subcontractors that we subcontract to are specialist. A direct worker, also known as general worker, working under the main-con needs to be taught how to do a particular trade. This will waste resources and precious time. Perhaps in other countries, the general workers they can specialise. However in Singapore, hard to find main contractors that can do specialised trades as there is not enough skilled workers in the market to begin with. In addition, as these sub-cons are specialist, they have the capability to do that particular trade. Hence, it will improve productivity.

7. **Do they have a core team of subcontractors whom they always work with? Why?**

Q7: Do you have a core team of subcontractors whom you always employ and if so, why do you keep rehiring them?

A7: No, we don't have as our core team of subcontractors always change. Why? It is simply due to capacity issues. At that point in time, even if we would like a particular sub-con to come and do a particular trade for us but they are bogged down by work from elsewhere, we will not employ their services as it presents a risk to us. Next, the boss of the subcontracting firm or the foreman is very important as a good boss or foreman can assemble a good team that is efficient in getting the job done. If the foreman or boss is not in firm A but in firm B at present, then we rather go for firm B. In addition, price is itself a very important factor. If the price the sub-con quote us is too high, we don't go for it. This is because these sub-cons they already have enough jobs to sustain themselves. Thus, pricing high is a way for them to avoid getting selected for the job even though they have been asked to tender.

We call for tender after we do a pre-qualification and we invite the respective shortlisted sub-cons to view the job information together with me on a one-to-one basis. Then we ask them to quote. From there, then we negotiate. Hence, we always compare the subcontractors, hence we don't have a core team of subcontractors.

8. **How do the main contractors help their subcontractors level up to enhance productivity, if any?**

Q8: How do you help your subcontractors level up to enhance productivity, if any?

A8: We don't help our sub-cons to level up to enhance productivity. The onus is on them to come propose to us, the main contractor, as they are the specialist for that particular trade. We are not. However, we do use an incentivisation initiative where the sub-contractors can get earlier progress payments if they complete their work earlier, unlike the stipulated date as agreed in the contract which might be later. We pay our sub-cons for the work done where the more they do, the earlier they get paid.

In addition, there is this new system called BIM, whose ideal purpose is to improve construction productivity, as recommended by BCA. However, I deemed that the adoption of BIM is very much too late. The BIM system is used by the Architect and his consultant team to input in data and create a model. However, in reality, coordination issues amongst all parties plagued the smooth adoption of BIM. Some of the parties are not even well-versed in BIM, so it defeats the purpose as the model is incomplete! Not only that, it is also very time wasting as most industry people are not familiar with such a system, especially the older generation who needs to spend even more time to grasp the system. Not easy! There is no such leisure time at the tender stage where you are only given 1 or at most 2 months to tender. For a complete model, it can take at least 3–6 months of work. Hence, where got enough time?

This is why BIM might not be widely adopted because of coordination issues where your main-contractor knows how to use BIM but not your subcontractors. As such, with regards to productivity improvement at the end of the day, it is at the main-con's expense whether to assist the sub-cons to improve productivity, where there is a price to be paid when we try to optimise productivity and cost.

Nonetheless, BCA does hold seminar conferences from time to time on how to improve productivity by introducing new methods of working and advanced machinery to boost productivity. Hopefully the sub-cons will attend.

9. **Is your firm also a subcontractor?**

Q9: Is your firm also a subcontractor?

A9: No. We are a main contractor.

10. **And if they are subcontractors, what and how would they want their main contractors to help them level up for construction productivity?**

Q10: So with regards to the subcontractor, what and how do you think the subcontractors want their main contractors to help them level up for construction productivity?

A10: For us, we will simply ask the sub-cons directly after quotation to negotiate and sell their innovative practices on how best they can complete the job on time via productive means such as using lesser workers within the same timeframe etc.

End of Verbatim Report 6

Annex 11: Interview Transcript 7 (Main Contractor I7)

Date of interview: Wednesday 28th Jan 2016.
Position of interviewee: Managing Director.

1. **Does your firm sub-contract works out to others?**

Q1: Does your firm sub-contract works out to other subcontractors?
A1: Yes. We do.

2. **What percentage of work does your firm typically sub-contract out to others?**

Q2: May I know what is the percentage of work that your firm typically sub-contract out to your subcontractors?
A2: It's about 80%.
Q3: Then the remaining 20% are works that are done in-house, right?
A3: Yes.
Q4: May I know what sort of works that is done in-house by your firm?
A4: We do structural works, architectural works such as wet trade (plastering) and tiling, M&E works and metal works. All these done in-house. No doubt we have our own workers but depending on the number of projects that we have undertaken, we will also sub-contract out if we have too many projects. For example, project A we sub-out, then project B we use our workers and do on our own. So we do it this way. Well, our firm is a subsidiary of a bigger parent firm.
Q5: So the parent company of your firm is based in China?
A5: Yes, and it is a state-owned enterprise.

3. **What types of work/trades do you typically sub-contract out to others?**

Q6: So what are the types of trades that you typically sub-contract out?
A6: Except for the trades mentioned earlier in question 2, we will sub-contract out all other trades.
Q7: So what type of projects does your firm usually bid?
A7: We bid for all kinds of project, including MRT projects.

L. Sui Pheng and S. Jia Zheng, *Construction Productivity in the Multilayer Subcontracting System*, Management in the Built Environment,
https://doi.org/10.1007/978-981-13-1831-3

4. **What guiding principles do you adopt to decide on (a) what, and (b) how much work to sub-contract out to others?**

Q8: So for your case, any guiding principles to determine what to subcontract and how much work to subcontract?

A8: First thing we look in a subcontractor is relevant track record. For example, they must have completed projects and have the experience in tiling works for some HDB projects etc. Secondly, their company resources, they must have sufficient resources.

Q9: Sufficient resources in terms of what aspect?

A9: In terms of finance, manpower in terms of number of available workers. For example, if a subcontractor only has 20 tilers, then we will not sub-out to this subcontractor as it is too small.

Q10: Any other guiding principles besides the ones you have mentioned which are very common in nature?

A10: Ok. I think we also need them to be BizSafe Level 3. MOM has also recommended to select sub-con that have a certain competency in workplace safety. This can help to improve the safety standard in our construction industry.

Q11: So is this more of a regulatory requirement from MOM?

A11: I would say it is more of an advisory.

5. **In the midst of a tightening labour market and increases in foreign worker levies which has resulted in a squeeze in foreign worker numbers, do you sub-contract more works out to others than ever before?**

Q12: Now under the midst of the tightening labour market like the reduction of MYEs etc., does your firm sub-contract even more works out than before?

A12: In that sense, yes. Some of our in-house workers is not as productive as those professional subcontractors. No doubt we can save more money when I sub-contract to my own guy. But at the end, it's the opportunity cost so end up we have to incur more MYEs. So might as well I sub-out to those subcontractors. For example, a tiling subcontractor company, their productivity higher, so they do not need so much MYEs. So for example, if I have my own workers to do tiling, I need 100 MYEs. If I sub-out, productivity higher thus no need so many workers, then need only 70 MYEs. So we will actually sub-out more.

Q13: So at the end of the day, it is the competency level and specialisation of subcontractors you are looking out for when you sub-out?

A13: Yes.

Q14: Just now you mention that you sub-out to your own workers. What do you mean by that?

A14: Oh, it's to my own subsidiary firm. We got different subsidiary companies. We got a company called LTS that does wet trade and tiling and structural works, then we got another company called Noszak that does metal works and steel structure works. Then, lastly we have Resources Great M&E that does M&E works. All of them are subsidiary firms of my company.

Q15: Then does your firm have their own general workers?

A15: Yes, we have.

Q16: So what is the size?

A16: We have about 2700 general workers. Then these workers are spilt amongst all other projects.

6. **Would sub-contracting out your firm's works to others affect your project construction productivity?**

Q17: Would sub-contracting your firm's work affect your construction productivity?

A17: Yes, it will. I would be more productive as the specialist subcontractors can have the expertise to do the job more efficiently. Thus, they can take a shorter time and even lesser workers to complete the job. Hence, it is more efficient to do so, hence resulting in higher productivity.

7. **Do they have a core team of subcontractors whom they always work with? Why?**

Q18: Do you have a core team of subcontractors whom you always employ?

A18: Yes. We do direct sub-out. This is because their productivity is higher, more skillful and have experience in their specialised trade. Have you ever heard of constructability score? So that's why we also adopt some of the advice given by the authorities in order to help our sub-cons. For example, the types of formwork will determine the C-score that we get. We use aluminium formwork instead of metal formwork. This is because it is lighter and results in higher C-score. B-score is mainly for the consultant design team. For formwork that has a higher C-score, it means more productive. So the formwork is precast, so we adopt more precast. So this is how sub-con improve their productivity. They (BCA) discouraged the traditional timber or metal formwork. So now contractors follow. For structural, we also use self-compacting concrete, so no need to compact. Thus, less workers are needed during concreting etc. So this C-score segment directly helps to improve productivity.

8. **How do the main contractors help their subcontractors level up to enhance productivity, if any?**

Q19: How do you help your subcontractors whom you always subcontract works to, assist them to level up in order to enhance productivity, if any?

A19: Yes. We do. As a main-con say structural work, material is supplied by us. So we take the decision to use aluminium formwork. Then, the sub-con will use aluminium formwork that will be more productive.

Q20: And usually that's the case in the industry where the main-con will tell the sub-con what measures to use to be more productive?

A20: **Yes**.

Q21: So usually, the sub-cons seldom propose to you work methods that can improve productivity right?

A21: Very rare, they are also not capable to do that. Their knowledge wise on productivity and technology such as equipment that can be used to improve productivity is not as good as that of main-cons.

9. **Is your firm also a subcontractor?**

   NIL
   Not applicable, firm is a main contractor.

10. **If your firm is also a subcontractor, what and how would you want your main contractors to help you level up your construction productivity?**

Q22: So has your core team of sub-contractors gave feedback to you, the main-con, on how you can help them to improve productivity? Do they even take this initiative to do so?

A22: Some simple measures such as they request to use concrete pump instead of using crane and bucket way to transport concrete to higher floors. But seldom they tell us what to do, it's us that recommend to them.

Q23: So as the main-con, you will also provide the equipment?

A23: Yes, we do. This is especially so if the equipment can improve productivity.

Q24: So I heard from other main-cons that there is a limit to mechanisation? Is that what you agree to as well?

A24: I think the limit they referring to is the grant amount that they can get from BCA. They must also see whether the machinery is suitable for the trade. Perhaps the limit can be due to size of the projects. Such as the self-levelling equipment for lean casting but this machine is more applicable to factory projects as it is bigger for the area. For residential projects, cannot use as the room size so small, how to use this machine? So that is what they mean by the limit.

Q25: Other main-cons also mentioned about some of the trades cannot use machinery at all as there is no such machinery even invented in the first place. What's your take?

A25: For example, tiling work. Even though there is research carried out such as using robots to lay tiles but still under R&D phase.

Additional Points

Q26: As a main-con, what is your take on construction productivity?

A26: Productivity is a trend and requirement to be more productive as we indirectly got no choice as it is a government objective as they are also directly decreasing MYEs to exert pressure on the industry to be more productive. But government agencies like BCA and HDB are also very helpful and advise us and educate us on new technologies that can help improve productivity. So as a main-con, I always tell my friends that we must be positive to all these changes and this is the trend. In fact, one of the issues is that there are too many foreign workers in Singapore and thus government is reducing the MYEs. Nonetheless, it is the Singaporeans who do not even want to work in this industry that is root cause of this problem. Hence, we have no choice but to engage foreigners but we also cannot afford to have too many foreigners. So this is how this productivity drive comes in. In fact, it is much cheaper to get foreigners but it is not sustainable. An example is China. Last time, 90% of the workers from China. Then before that, it is from Thailand and Malaysia. But right now, there is a decreasing number of China workers coming, all because of the one-child policy. So current china construction worker does is not born under the one-child policy period. The parents sure pamper their only child. So in the next 10 years, the foreign workers from China will get lesser, hence government is sourcing more foreign workers from other countries like Myanmar and Vietnam. However, these new foreign workers from Vietnam and Myanmar skill very low, unlike the Malaysians, who have the best skill amongst all. Sadly, Malaysians gradually don't want to come here to work, then we have to source workers from China. China workers' skills not too bad. Then now, we have to source workers from countries like Myanmar and Vietnam, their skill level is getting poorer. So might as well we try to increase productivity, use less workers and then can maintain the quality of workmanship. This will be more sustainable in the long term.

Q27: Besides this, what else do you see in terms of subcontracting practices that can have an impact on productivity?

A27: Cannot have too many layers of subcontracting. If too many layers then got problem already. For example, I am the main-con and I sub-out tiling works to company A. Then company A might sub-out to Companies C and D. Hence, got 3–4 layers. Then at most we try to choose direct sub and we don't expect them to have a few more layers under them. We expect them to have their own team of workers under them and doing the job by themselves. Hence, more consistent on quality and productivity. So skill level, productivity level will become higher.

Q28: You also previously mentioned that you will do direct sub-contracting to ensure QC. Is this direct sub-contracting also known as domestic sub-contracting?

A28: Yes, it is. So that's why the subsidiary firms like us under our parent company are all our domestic subcontractors and thus it is more reliable.

End of Verbatim Report 7

Annex 12: Interview Transcript 8 (Main Contractor I8)

Date of interview: 27th Jan 2016.
Position of Interviewee: Managing Director.

1. **Does your firm sub-contract works out to others?**

Q1: Does your firm sub-contract works out to the subcontractors?
A1: Yes. But only for specialist trades like M&E works like Electrical works, etc. or building works like aluminium and glazing works, etc.

2. **What percentage of work does your firm typically sub-contract out to others?**

Q2: What is the percentage of works your firm typically sub-contract out?
A2: About 30–40%.
Q3: Typically, most of the other main contractors sub-out about 80–90% of their work to subcontractors. So this means that you will do at least 60–70% of the work in-house. Why is this so?
A3: Basically for us, we are an SME (Small and Medium Enterprise) firm. So when I tender for jobs I usually look for jobs that my own workforce can handle, except for those specialist trade like tiling works, electrical, ACMV, lifts etc. We have a subsidiary firm that does iron works and metal works. For that subsidiary firm of mine, they are doing more metal works like structural steel etc.

3. **What types of work/trades do you typically sub-contract out to others?**

Q4: So what are the types of trades that you typically sub-contract out?
A4: We sub-out building works like aluminium and glazing works, tiling, etc. and M&E works like Electrical, ACMV, Sanitary, Plumbing and Lift work.

4. **What guiding principles do you adopt to decide on (a) what, and (b) how much work to sub-contract out to others?**

Q5: What are your guiding principles to determine what to subcontract and how much work to subcontract?

L. Sui Pheng and S. Jia Zheng, *Construction Productivity in the Multilayer Subcontracting System*, Management in the Built Environment,
https://doi.org/10.1007/978-981-13-1831-3

A5: The principles we adopt are:

- The number of projects on hand
- Skillset of the workforce
- Market condition etc.

Normally for us if we are going to subcontract out is based on the number of projects that I have on hand and whether the job I get can have a good profit if I sub-out or if I do it in-house. Fundamentally, it is ultimately still about the profitability, workmanship and quality. Even if you are a big firm and you don't look into all these, there will not be any more repeated clients already. As we have been in the market for a very long time, so to us workmanship is very important and therefore, we have a lot of repeated customers and they are also big firms. So if you do for them well for the first project, workmanship is good all the time etc., these big firms will call you up again for future jobs.

Q6: You must have quite a substantial number of workers in your firm since you are doing most of the works in-house?

A6: Not really, we have only 30–50 workers and they consist of general and skilled workers. So for us it is not so much of the turnover or revenue but more on the profits and on the management side. We will selectively choose the jobs that we want to do, so that we can maximise our resources and at the same time, we make sure that we get maximum profit out of it.

5. **In the midst of a tightening labour market and increases in foreign worker levies which has resulted in a squeeze in foreign worker numbers, do you sub-contract more works out to others than ever before?**

Q7: Now under the midst of the tightening labour market like the reduction of foreign labour quota etc., do you sub-contract even more works out than before?

A7: Yes and No. Yes is due to the insufficient manpower to carry out the number of projects on hand. No is due to the ability of our firm to tender for projects within the company resources so as to make full use of the company resources. Ultimately, it comes down to the market condition.

6. **Would sub-contracting out your firm's works to others affect your project construction productivity?**

Q8: Would sub-contracting your firm's work affect your construction productivity?

A8: Yes and no as it is dependent on the responsibility of the subcontractor which is determined by the working relationship, jobs on hand, manpower, financial status of the company etc.

Q9: From a main contractor's perspective, what is your definition of construction productivity?

A9: Basically construction productivity is to use minimum workforce to get maximum returns, finish the job on time and using the most effective construction method.

Q10: Some other main contractors mentioned that there is a limit to how productive you can be in terms of mechanisation of work on site. Is that a sentiment you share with them as well?

A10: Definitely. So that's why the first principle when you tender for job is whether you can achieve the outcome that you want to get. We are not looking at mass market tendering of getting jobs but selectively.

Q11: So being selective of what sort of job to tender ensures productivity?

A11: Yes. Not only does it ensures productivity, it also ensures quality workmanship as it is done by us in-house. This, we can guarantee that. If we sub-out to sub-cons, we will never know. It may be that outside they have other jobs and they need to pull their man out, or suddenly they are working for someone else and then the man-con don't pay them. Financially they suffer and it indirectly affect all the other jobs they have. So these are all a chain reaction. We need to be concern and be prudent on all this issues.

7. **Do they have a core team of subcontractors whom they always work with? Why?**

Q12: Do you have a core team of subcontractors whom you always employ and if so, why do you keep rehiring them?

A12: Yes. This is because we are aware of their workmanship, resources available, financial status, etc.

8. **How do the main contractors help their subcontractors to level up to enhance productivity, if any?**

Q13: How do you help your subcontractors to level up to enhance productivity, if any?

A13: We will request that they send their workers for relevant skill upgrade, etc. You are aware nowadays that to be able to operate those mobile-elevated platforms, scissors lift, boom-lift, it require certificates as proof of competency. So as main-cons, when we tender for jobs, it's obvious that the client requires this as a requirement in the contract. This is a very common mass market thing and we actually inform our subcontractors that they should get all these. They shouldn't expect the main-cons to provide the operators and they only provide the workers. They shouldn't be telling us that they are not qualified or competent simply because they do not have the necessary certifications. I think these are all the basic certifications that all sub-cons should have when they step into any site. This is a requirement to have the necessary certifications where they at least can come into the site and improve their productivity. Productivity is when your workers is always doing the job and then from there they know what are the pros and cons due to experience gained from site. This also help them be familiar with the work process and eventually improves productivity.

Q14: So the onus is on the main contractors to inform the subcontractors to go for such upgrading to improve their skills and therefore improves productivity?

A14: It also depends on the subcontractors as well. If you have those that are longer established, they will go these upgrading courses and certifications by themselves.

9. **Is your firm also a subcontractor?**

Q15: Is your firm also a subcontractor?
A15: No. We are a main contractor.

10. **And if they are subcontractors, what and how would they want their main contractors to help them level up for construction productivity?**

Q16: So with regards to the subcontractor, what and how do you think the subcontractors want you, as their main contractors to help them level up for construction productivity?

A16: Normally the sub-cons will do the necessary skills upgrading and certification etc. by themselves. My subcontractors are regular sub-cons. They have been with us for a very long time, so we have this common understanding. In addition, we select our subcontractors very stringently. We have to see their performance and track record. Our company has been around for very long and so if a sub-con that we anyhow select cannot perform, then in the end it's still us. So that's why we select our sub-cons quite carefully.

Additional Points

Q17: So currently in Singapore, the construction industry isn't doing very well as main contractors are struggling to "survive" as well. What is the root cause of it?

A17: Basically, I think it is the general economy of the whole world that's affecting the industry as well.

Q18: So in terms of the current practice of subcontracting being carried out in Singapore, do you think it is causing a problem for both main-cons and sub-cons to become more productive?

A18: Basically, now the economy is no good and you as the main contractor, you sub-out all your work your margin won't be very good. Your profit is not good. When you sub-out to people, how much can you mark up? This is because your sub-cons have already mark-up all they need to mark-up already, so your margin will be very thin. Secondly, in a way you are just doing project management. Because you are just like an agent; you get the job, then you divide into phases. Besides that, you don't get that much control as well. Firstly, in terms of workmanship, subcontractors are also tie up by their budget as well. So let's say they tender to you at this price, which is a lousy price, then you keep asking them to do their work properly, then after once or twice, they will stop already. They will complete the work but give a shoddy job, all because you got them at a cheap rate as they only tender for this amount, thus we cannot expect them to give you above this amount. They also won't do it.

Q19: Are there any other common problems with regards to the subcontracting practices in Singapore?

A19: Maybe it could be actually wanting to establish their name in the market. This is because some contractors they just do a job and then they just close it or if they cannot do, then they just close it. It is not for the long term prospect. If you are looking at the long term, then it is different. For sub-cons and main-cons it's the same. It all depends on the boss's mind-set as to whether he wants to take the business for 20 or 30 years, not just 1–2 years only.

Q20: So it's all because most of the sub-cons they fold pretty fast I suppose?

A20: Yes, precisely.

Q21: What about BCA's drive to boost productivity in our sector via different initiatives which they have recently launched; do you think it is effective?

A21: It's effective if you have jobs. Let's say in terms of mechanisation. Even if they subsidised for you 40% to purchase a new machinery that can improve productivity but in the end you can't get any jobs and yet must still store your machine, it doesn't make sense.

Q22: So it's also boils down to availability of jobs to improve productivity, am I correct?

A22: Yes, it depends on the market conditions. Productivity also depends on the market conditions. Let's say if there is a recession, nobody will think of productivity right? It's survival for firms now.

End of Verbatim Report 8

Annex 13: Interview Transcript 9 (Main Contractor I9)

Date of interview: Saturday 30th Jan 2016.
Position of Interviewee: Project Manager.

1. **Does your firm sub-contract works out to others?**

Q1: Does your firm sub-contract works out to other subcontractors?

A1: Yes. We do that of course. We do that because they don't have the expertise for some of the trade, they want efficiency and they want to reduce their cost. Frankly speaking, the skill of the workers keeps dropping and that's why when compare with workers from around 20–30 years ago, the skill really dropped a lot. Last time when we are supervisor, site engineer or site manager, the skill level that can be achieved by our workers then easily exceed our expectations of a good quality job and they can even take over the boss's job when needed. Today, our workers cannot make it.

Q2: So what attributes to the sharp decline in skills of the workers?

A2: Mainly because last time we still have local workers or some Malaysian staff that have more loyalty to the company and to the country, they did not have those "hit-and-run" kind of attitude like what we have today as they don't want to stay in the industry for long and they really have the passion to pick up the kind of skill to learn and improve themselves. No doubt BCA have a lot of courses to train the workers. It is useful but how you apply the knowledge to the practical job takes time.

Q3: So you are saying that not many people want to stay in this industry for long?

A3: Yes, so that's why the skill and competency level is not there, hence the constant decline of skilled workers in this industry. So now sub-cons focused on their specialised trade and do it well. Unlike last time, we as the main-con we take up the job and do it without even the need to sub-out to sub-con.

L. Sui Pheng and S. Jia Zheng, *Construction Productivity in the Multilayer Subcontracting System*, Management in the Built Environment,
https://doi.org/10.1007/978-981-13-1831-3

2. **What percentage of work does your firm typically sub-contract out to others?**

Q4: May I know what is the percentage of work that your firm typically sub-contract out to your subcontractors?

A4: It's 100%. What we have is some of the general workers to do touch-up, move rubbish and bulky items or even doing some preliminary works like pump water or electrical works. So these general skills don't need to be too skilled.

Q5: So the general workers are for general maintenance of this site office?

A5: Yes.

Q6: So how many general workers do you have, say in this site?

A6: Roughly around 20–30 workers easily for this kind of site. At the peak, we have around 40 general workers.

3. **What types of work/trades do you typically sub-contract out to others?**

Q7: So I presume that you sub-out all sorts of trade since it's 100% subcontracted out?

A7: Yes, indeed. Everything.

4. **What guiding principles do you adopt to decide on (a) what, and (b) how much work to sub-contract out to others?**

Q8: So for your case, any guiding principles to determine what to subcontract and how much work to subcontract?

A8: Initially, we sub-out in such a way where we have a few types of arrangement, like in big packages like formwork and rebar for main steel structural works, we put in one package, then architectural works like wet work and dry work in one package. Normally, if the sub-con can do it like one bundle in all these packages, then they are more or less like main-con. They are not picking up such a job. So why should I sub to you when I rather become his main-con and do the work on a smaller scale. So actually, we go by trade to trade.

Q9: So you prefer to focus on each individual trade and sub-out?

A9: Yes, but the structure wise like formwork, rebar, concrete we always sub-out to one sub-con to do. In the earliest stage, we also break it into the 3 to do it, but we find it quite messy and the coordination is difficult and wastage is high.

Q10: So your guiding principle here is that when you sub-out your works, you want to ensure that your subcontractors can do things in a bundle?

A10: Not do in a bundle, we try not to do it as it can be a headache as well. What I am saying is that those sub-cons that can do it in a bundle, they are supposed to be very high skilled. But some of them when they bundle and when they encounter a trade which they do not know, they subcontract out to another party (sub-sub-con), so why should my sub-con get the job then subcontract it out again? Then every time when there is a problem, they will come back and look for us. I rather I directly deal with them the problem than the sub-sub-con inform the main-con and then the main-con then come and inform me.

Q11: So it's like direct contracting, thus easier to manage?

A11: Yes, that's right.

5. **In the midst of a tightening labour market and increases in foreign worker levies which has resulted in a squeeze in foreign worker numbers, do you sub-contract more works out to others than ever before?**

Q12: Now under the midst of the tightening labour market like the reduction of MYEs etc., does your firm sub-contract even more works out than before?

A12: No difference at all. Even in the past and even now, we have always been subcontracting out everything.

Q13: So even before the labour crunch occurs, your firm has always been subcontracting out 100% of your works?

A13: Yes. Frankly speaking those that are able to do the works on their own are mainly the smaller scale contractors as they are more cost effective and efficient. An example like these small-scale contractors that are building one pair of Terrace house or bungalow and the contract sum is less than a million dollars and you only have less than 10 workers per day. Then how can you sub it out in the present market? It is very hard also. So you try to train the worker to multi-task by getting a more skilful kepala or foreman to supervise them. They are a bit slow but you still get the job done. The quantum too small already and people would not want to mobilise. How can you sub it out? Nowadays, not like last time doing sub-con, do a job requires little knowledge. Now it's not just bringing their tools to work, they also need to bring the pen to work. They everyday also need to fill up those safety forms and other administration paperwork, then you know all these workers everyday do work on site, you think they know how to read? They don't even know how to read, then how can they fill up a form? So the contractors have no choice but to engage those Bangladeshis or Indians who can read English and some are quite highly educated. They can fill up a form. You get those China workers to fill up a form, they can't as they can only read Chinese. So some of the china contractors they also engage some of these Bangladeshi or Indian workers who can read and write for some of the safety related positions like signalman, rigger man, safety supervisor, WSHA supervisor etc. They are the ones that can fill up the form and then can attend meetings and thus can still communicate with them.

6. **Would sub-contracting out your firm's works to others affect your project construction productivity?**

Q14: Would sub-contracting your firm's work affect your construction productivity?

A14: No, I don't feel any difference. But when we commit to such a job, then we should prepare the resources to do this job. If not, better not do the job. But sometimes got a lot of cork-up, such as not enough worker when their permit expires or they are unable to get new workers to come in and do the job. Such a situation also have. Another situation quite common such as site issues where the workers need to be urgently deployed to another site and cannot comeback on time etc. Those kind of situation also have.

Q15: So what is your view on construction productivity?

A15: Usually if we sub-out, we will assess their previous performance and if they hit our requirement, we will sub it out to them. The new sub-cons that are not familiar with us, such as when the consultants appointed them or they are the only specialist around and we still have to work with them, such things are hard to control. All those specialised contractor, with regards to productivity, they are in this trade for many years already. Thus, they should have the skill and resources to do their trades. For example, plastering trade I cannot just get my own general workers and do this job because they don't have the necessary expertise to do plastering.

Q16: So in other words, the efficiency of the job getting done should increase when you sub-out to the subcontractors who specialise in this trade?

A16: Yes, because if you do everything on your own, you will have to do a lot of investment. You have to buy your own equipment, tools and pay your skilled workers. In construction, there are many different components. You have architecture (Wet trade and dry-trade), structural etc. So there are not many workers that can command so many skills. Even if they command that much skill, their productivity also cannot be so high. So you cannot pay them so high and expect them to be very productive.

Q17: So you are saying that even though these workers may be multi-skilled but they are the "jack of all trades, master of none" type of workers? Am I correct?

A17: Yes, however these types of workers they may be multi-skilled but this kind of workers are good when you are doing those small-scale projects like building terrace houses or bungalows. Hence, with such workers, you no need to have a headache on the manpower issues and the need to look for workers that specialised in a particular trade which is needed in your project. So they have these kind of specialist contractors, where their cost is more competitive and more efficient. These firms can pay a little more and keep them as their own direct workers. Then, you can do all sorts of works.

For example, in Australia, if the house is not more than 2 storey high, there is no need for Architect's submission. The builders can do the submission. So they already have the module house readily available for clients to choose. After the client has chosen a particular module house, the builders just send 6 workers to build the entire house from scratch; From foundation all the way to completion and key handover. These 6 workers handle everything. All the coordination no headache at all. They know how to read drawing and calculate. Singapore so far do not have such a skilled worker.

Q18: Why do you think Singapore don't have such "multi-skilled workers"?

A18: These can be many factors such as lack of approval and recognition from developer, builder, designer and authorities—Mind-set. In Singapore, the status of the contractor is extremely low. However, in overseas, the status of the contractor is on par with the rest of the professions. Even a brick-layer is very proud like one of the professional workers. For us, if we tell the bricklayer to stand beside the Architect and say that they are of equal standing, you get ridiculed and be dissed of instantly. Hence, the kind of working attitude is different here.

Q19: So it boils down to the mind-set of the industry, especially in Asian societies?

A19: Yes, indeed.

7. **Do they have a core team of subcontractors whom they always work with? Why?**

Q20: Do you have a core team of subcontractors whom you always employ?

A20: Yes, but sometimes also depend on their schedule as well. Sometimes the sub-cons have too much of a commitment and they cannot do for you. You can't force them as well as you don't get any advantage. This is because they also lack of the resources to do it for you. It also depend on the scale of the contract, how big it is, enough manpower to work from one job to the next without disruptions and also how they schedule their works. In addition, they must also see whether they can manage their manpower. Nowadays, the levy, labour quarters and transport all very high. They cannot afford so many of the spare workers hang around and do nothing. So we also understand their situation. Sometimes, when the main contractors when they get a very huge job, they sub-out the works to 2 or 3 sub-contractors. They cannot just let one sub-contractor handle the entire job. Of course it's best to have 1 sub-contractor to do the work, but when there are a few sub-contractors working together, then you are likely to have interface and coordination issues etc. Then you will have a headache.

Q21: So when the scale of the work is too big for even your core team of sub-cons to handle and must further sub-out, then how are you going to select suitable sub-cons?

A21: We will assess their performance and history and also check with the market how's their performance etc. We also check their offering price and their stated terms and conditions when they want to do up the job. Every sub-con they have different terms and conditions when they work with us.

Q22: So these core team of sub-cons is mainly for the reliability and ease of communication and working with them, am I right?

A22: Yes.

8. **How do the main contractors help their sub-contractors level up to enhance productivity, if any?**

Q23: How do you help your subcontractors to level up in order to enhance productivity, if any?

A23: Basically, they have their own workers and they themselves should send the workers to go for training as this is the norm. Whereas those under our own payroll, we will send them for training. For sub-cons, not so likely. Of course we did invite specialist to come and do some demonstration, then we also invite our sub-con to come and take a look tom enhance their knowledge.

Q24: So when there is a new way of working or some new technology, you will actually share it with your core team of sub-cons?

A24: Yes.

9. **Is your firm also a subcontractor?**

Q25: So your firm is definitely not a subcontractor?

A25: Sometimes, we also become a subcontractor when we have a special project. One example would be when we do a jet engine test facility for a client where they call tender for test engine cell. So the test cell alone which they want to build, they sub-out to the mechanical engineering firms. So one of our counterpart firm, they got the job. However, some of the civil engineering part of the job, they sub it out to us. Like those structural works, architectural work and basic M&E works. Only the specialist equipment is installed by them. So they become the main-con, we become the sub-con. This is a very special arrangement. Just like power gird works. Sometimes when tender, they look for electrical contractor. So the electrical contractor becomes the main contractor but the structural and architectural portion they can't do it such as laying cables etc. Hence, they sub-out to civil contractor where they can do the structural and architectural works and hence they become a sub-contractor. So these are all special arrangement.

10. **If your firm is also a subcontractor, what and how would you want your main contractors to help you level up your construction productivity?**

Q26: So has your sub-contractors gave feedback to you, the main-con, on how you can help them to improve productivity? Do they even take this initiative to do so?

A26: We will state clearly and agree in an arrangement where the main contractor will provide a few main equipment for the subcontractors to use such as the concrete pump, tower crane etc. The sub-con will provide minor tools for themselves. However, if the sub-cons want to supply their own equipment, by all means they can do so.

Additional Points

Q27: In general, what's your opinion on the subcontracting practices that is occurring in our construction industry? Are there many layers of subcontracting going on?

A27: Not really though. If it is planned properly, then there shouldn't be too many subcontracting layers. Unless the job scope is too big, then the sub-con unable to finish the job, then we as the main-con can consider dividing the job into a few parts and then sub-out to different sub-cons. However, if you have too many sub-cons, then it may affect work quality as sometimes the sub-cons they don't really care much.

Q28: Well, are there any instances where the sub-con further subcontract out their works to another sub-sub-con and so on, causing increasing levels of subcontracting to occur?

A28: Yes, there are. These are actually not sub-cons. They are called agents. They don't have anything and they just use their name to get jobs only. Sometimes, we did encounter all these but it should not be encouraged, you know what I mean… Usually, such things will result in lots of headache for us.

Q29: So if you find out any of your subcontractors that do this, you will 'terminate' the contract with them?

A29: So that's why we have to look at the track record to see whether these subcontractors are genuine. If the subcontracting firm is just an appearing agent, then there will be lots of "behind the scenes" going on which I don't want to comment. Sometimes the GM or the bosses of these subcontracting firms they do something like this, but it isn't too good. It's worse if the agent doesn't know this trade and you talk to him and then he relies heavily on his so called "sub-con" So what for you want to talk to him?

Q30: Do you have any views on the state of the construction industry in Singapore pertaining to productivity?

A30: It's because of the poor attitude in this industry, people not willing to stay long in this industry. Hence, this creates a big gap. Singaporeans don't really want to pick up the skills and do hands-on. Locals always want to define their scope well, so once you do that, you wouldn't want to learn more right? They can't do hands-on and get themselves into the process of solving problem. As such, when there is a similar problem that happens in future, cannot tackle it. Unfortunately, the mind-set now is so scope defined, most people in this industry don't want to learn anything else that is not in their job scope. As such, it is very hard for productivity to take place, unlike last time, where for example, even a typical site engineer can take over the role as a PM easily as people in the past are exposed to all sorts of work that is not even in their job scope and they do so willingly, even it's the PM's job. This therefore ensures productivity does not decline. Now, times have changed. It's hard times…

End of Verbatim Report 9

Annex 14: Interview Transcript 10 (Main Contractor I10)

Date of interview: Sunday 31st Jan 2016.
Position of Interviewee: General Manager.

1. **Does your firm sub-sub-contract works out to others?**

Q1: Does your firm sub-sub-contract works out to other subcontractors?
A1: Yes, we do.

2. **What percentage of work does your firm typically sub-sub-contract out to others?**

Q2: May I know what is the percentage of work that your firm typically sub-contract out to your subcontractors?
A2: Around 5–10% of my work.
Q3: So most of your work 90–95% done in house, right?
A3: Yes.

3. **What types of work/trades do you typically sub-sub-contract out to others?**

Q4: What sort of trades did you sub-sub-contract out to your sub-sub-contractors?
A4: Before we decide whether to sub-out, we see whether we have the capability to do it ourselves, e.g. we do toilet cubicle and sometimes customers have weird requests, they want glass on their toilet cubicle. And we don't have the machine that cut the glass. So we sub-out this part of the work to my other subcontractor and then they will cut the glass for us and then they will proceed as usual. We take the glass back, then supply and install. So long as it is out of our capability to do such a thing, we sub-out.
Q5: So you mainly sub-out because you don't have the necessary expertise?
A5: That's a main reason. There are of course other reasons. Another main reason why we sub-out is because when we are doing construction, sometimes there are a lot of job fluctuations. During the peak period, there are a lot of jobs coming in

L. Sui Pheng and S. Jia Zheng, *Construction Productivity in the Multilayer Subcontracting System*, Management in the Built Environment,
https://doi.org/10.1007/978-981-13-1831-3

and you need to have a lot of manpower. Or else you cannot complete the jobs on time. Usually we have to stick to a certain schedule. So we cannot overburden ourselves by employing too much people. If we do it, fixed pay fixed cost even when there is no job. Hence we make a loss. What we do is that during peak period, we sub-out part of our work, so we don't have to pay for the workers once there is no work, so this is cost savings.

Q6: So this is cost effective in that sense?

A6: Yes.

Q7: Any other works you sub-out?

A7: So far these are the 2 main ones. One is cutting glass and the other is to save cost by sub-out other works.

4. **What guiding principles do you adopt to decide on (a) what, and (b) how much work to sub-sub-contract out to others?**

Q8: So for your case, any guiding principles to determine what to sub-sub-contract and how much work to sub-sub-contract?

A8: I think we look at profit. So if we decide that if we sub-out, we can have better profit margin then we will usually do it, that's usually the guiding principle.

Q9: Any other guiding principles?

A9: Usually we also see how much workload we are already handling. So let's say if our people handle too much jobs and therefore super-tired already. It is here where our workers' salary can be even higher than our office staff, all because of OT (Overtime). So we can see that they are already working too much already and when that happens, we also will sub-out our works.

5. **In the midst of a tightening labour market and increases in foreign worker levies which has resulted in a squeeze in foreign worker numbers, do you sub-sub-contract more works out to others than ever before?**

Q10: Now under the midst of the tightening labour market like the reduction of MYEs etc., does your firm sub-sub-contract even more works out than before?

A10: That's a very tricky question. As a sub-con, we are not a huge company. We don't have lots of manpower like draftsman to manage our work. So when we sub-out, we usually face a lot of coordination problem. Let me give you an example: When we want another company to do work for us, sometimes they require us to produce shop drawings or something specific, things that require a lot of time to do and we have to spend manpower to coordinate with them just so that they can coordinate with us. When we have lots of layers of sub-con, it starts to get very hard to coordinate and there will be slower communication, leading to lower efficiency, worse still higher chances of miscommunication, confusion which can screw up the project. This is especially so when there is a cork-up and all parties point fingers at each other. It is an ugly scene.

6. **Would sub-sub-contracting out your firm's works to others affect your project construction productivity? How do you ensure your firm stays productive?**

Q11: Would sub-contracting your firm's work out affect your construction productivity? Any measures your firm has adopted to continuously stay productive?

A11: Yes. Our sub-sub-cons are specialised in their field of work and it will be more efficient for them to do the job as it can be done more quickly and with less manpower. In this industry, if you are the sub-con, it is still best for you to specialise as you can pull your resources together to ensure that your resources are more fully utilised, have lesser wastage and you can concentrate your efforts on making sure you excel in a particular trade. Once word gets out into the industry that you are good in a particular trade, you will end up receiving many jobs. That's how you survive as a sub-con in Singapore's cut throat industry.

7. **Do you have a core team of sub-subcontractors whom you always work with? Why?**

Q12: Do you have a core team of sub-subcontractors whom you always employ?

A12: Yes we have. We have already known them for quite a few years already as we are a family business and we are familiar with their working styles, how easy they are to work with, not demanding, thus it is easier to coordinate with them. For example, my core team of sub-sub-cons they no need me to pay them down payment first, they just do finish the job properly and then I come down collect and then pay them. No qualms or disagreement. All easy-going, hardly calculative. As such, this results in lesser miscommunication and errors occurring, which enables us to deliver our work more efficiently and on-time to our clients.

Q13: What happens if your core team of sub-cons are unable to take up more jobs from you? Do you stop sub-sub-contracting out your works?

A13: If their workload is to their maximum already and yet we still have jobs, we have no choice but to look for other sub-sub-cons. This poses a risk as we are unfamiliar with them. Hence, we have to scrutinise their track record, their expertise and experience and also their financial standing. If possible, we will try to hear from industry sources as to the reliability of the sub-sub-con we are hiring.

8. **How do you help your sub-subcontractors to enhance their construction productivity, if any?**

Q14: How do you help your sub-subcontractors to level up in order to enhance productivity, if any?

A14: We usually don't help our sub-sub-cons to improve their construction productivity as they are experts in their trade and shouldn't require any other expertise at all. In fact, so long as they can complete the job properly with quality work, not shoddy work and on time, I am more than happy.

9. **How does your firm value-add to your main contractors whom you work with?**

Q15: How does your firm try to value-add to your main contractors whom you associate with?

A15: We will definitely want to deliver our work on time and definitely that will value-add to my main contractors. However, we are aware that we must not give the main-con opportunities to deduct our down payment and worse still, our progress payments. The industry is bad enough and what I, as a sub-con asked for is just payment on time.

10. **If your firm is a subcontractor, what and how would you want your main contractors to help you level up your construction productivity?**

Q16: As a subcontractor, how would you want your main-contractor to help improve your construction productivity?

A16: I have a lot of frustrations on this matter. Firstly, the main-cons ought to be more understanding and less demanding. I say this is because some main-cons are unreasonable say in the length of time given to complete the job. Some of their time frame to do the job is just so ridiculous, it's way too short. Sometimes because of unforeseen circumstances like bad weather, haze, or even shortage of supply of materials, hence difficulty in ensuring that these material comes on time. It can also be that the main contractor's RTO come and inspect that there is scratches at one tiny corner of the wooden panel, so the work is unacceptable, must redo the entire panel. Hence, we incur additional cost to change and re-install a new panel and yet we are not reimbursed. Besides that, we still must complete the job on time. Sometimes, we also need to adhere to the client's demands directly as the main-cons referred them to us.

Secondly is payment. It is ridiculous for some main-cons not to pay their sub-cons in full even though they have completed the job on time and they always drag the progress payments etc. So we have actually blacklisted some of the main-cons in our list and we vowed not to work for them. Hence, these are the 2 most fundamental issues that I faced as a sub-con as we are sometimes at the mercy of the main-cons. How much we can survive depends on how much the main-cons are willing to pay us on time and how difficult they want to make our life be.

Additional Points

Q17: In general, what's your opinion on the subcontracting practices that is occurring in our construction industry? Are there many layers of subcontracting going on?

A17: In addition, the issue with our industry is too many layers of subcontracting. It has an overall bad effect on the industry. Especially now when times are difficult, jobs are hard to get, then main-cons will have to undercut/ do predatory pricing in order to get the job just to survive and tide through these tough times. Then because too many layers of subcontracting, then for us sub-cons will suffer as our progress payment will also be suppressed by the main-con. So the subcontractors that are at the most bottom of the subcontracting layers suffer the most and earns the least. Hence, everybody in the industry is struggling as well. To make matters worse, some of the main-cons don't pay us in full for a particular job that we have done already and choose to hold back our payment. Like for example, we still have 2–3% of our jobs that we undertake for the main-con, they still have yet to pay us and it has already been almost a year. I don't even see them being able to repay us the money.

Q18: Do you have any views on the state of the construction industry in Singapore pertaining to productivity?

A18: There is essentially a huge flaw in our subcontracting system here in Singapore. Because of the multiple layers of sub-con being done, it gets difficult to handle especially when things go wrong. It is hard to pinpoint the root cause of the fault from whichever party. This boils down to the challenge of communication and coordination that has plagued this industry for decades. Sometimes when the job scope, description and roles are not clearly define and then passed down the layers clearly, confusions arises. It is not like some other industry like IT where almost everyone knows their roles and responsibilities well and can be easily contacted without having to ascertain and find out the root cause of the fault when problems occur. Hence, having multiple layers of subcontractors hampers productivity and I try to avoid this.

End of Verbatim Report 10

Annex 15: Interview Transcript 11 (Main Contractor I11)

Date of interview: Monday 1st Feb 2016.
Position of Interviewee: Contracts Manager.

1. **Does your firm sub-sub-contract works out to others?**

Q1: Does your firm sub-contract works out to other subcontractors?
A1: Yes, we do.

2. **What percentage of work does your firm typically sub-contract out to others?**

Q2: May I know what is the percentage of work that your firm typically sub-contract out to your subcontractors?
A2: Around 95% of our works we have.

3. **What types of work/trades do you typically sub-contract out to others?**

Q3: So what are the types of trades that you typically sub-contract out?
A3: We subcontract almost all trades are sub-contract out mainly piling works, RC works, external works, metal and aluminium works, architectural works and M&E and specialist works.

4. **What guiding principles do you adopt to decide on (a) what, and (b) how much work to sub-contract out to others?**

Q4: What are your guiding principles to determine what to subcontract and how much work to subcontract?
A4: Well generally, our guiding principles are determined by 3 factors mainly; Relationship with the sub-contractor, how competitive is the pricing and their track records. As to how much work to subcontract out, it really depends on the capability and quality of work done by the subcontractor.

L. Sui Pheng and S. Jia Zheng, *Construction Productivity in the Multilayer Subcontracting System*, Management in the Built Environment,
https://doi.org/10.1007/978-981-13-1831-3

5. **In the midst of a tightening labour market and increases in foreign worker levies which has resulted in a squeeze in foreign worker numbers, do you sub-contract more works out to others than ever before?**

Q5: Now under the midst of the tightening labour market like the reduction of foreign labour quota etc., do you sub-contract even more works out than before?

A5: We still continue to award out sub-contract as per normal. In general, our sub-contracted works are still traditionally awarded as sub-contract work package.

6. **Would sub-contracting out your firm's works to others affect your project construction productivity?**

Q6: Would sub-contracting your firm's work affect your construction productivity?

A6: Yes, it will. The impact of project construction productivity will depend of quality of our subcontractor.

7. **Do they have a core team of subcontractors whom they always work with? Why?**

Q7: Do you have a core team of subcontractors whom you always employ and if so, why do you keep rehiring them?

A7: Yes, we do have one core subcontractor which we work together for many years and in many projects together. Thus this enables the reliability and standard of workmanship to be assured and achieved.

8. **How do the main contractors help their subcontractors to level up to enhance productivity, if any?**

Q8: How do you help your core subcontractor to level up to enhance productivity, if any?

A8: We usually conduct frequent dialogues and discussion with our core subcontractor. In addition, we also provide trainings and sharing sessions with them to ensure their skills remain relevant with the changing times and in line with government regulations.

9. **Is your firm also a subcontractor?**

Q9: Is your firm also a subcontractor?

A9: No. We are a main contractor.

10. **And if they are subcontractors, what and how would they want their main contractors to help them level up for construction productivity?**

Q10: So with regards to the subcontractor, what and how do you think the subcontractors want you, as their main contractors to help them level up for construction productivity?

A10: The services of our core subcontractor are regularly employed by us. They have been with us for a very long time, so we have this common understanding. In addition, we select our subcontractors very stringently. As such, our subcontractor will do the necessary skills upgrading and certification etc. by themselves.

End of Verbatim Report 11

Annex 16: Interview Transcript 12 (Subcontractor I12)

Date of interview: Tuesday 26th Jan 2016.
Position of Interviewee: Manager.

1. **Does your firm sub-sub-contract works out to others?**

Q1: Does your firm sub-sub-contract works out to other subcontractors?
A1: Yes, we do partially. Some of our works we combine with our sub-sub-con.
Q2: So what is the nature of your business?
A2: It is electrical installations.

2. **What percentage of work does your firm typically sub-sub-contract out to others?**

Q3: May I know what is the percentage of work that your firm typically sub-contract out to your subcontractors?
A3: We have our own manpower. Just in case we are unable to cope due to the period of time and the nature of the works. So these are the 2 criterion we look out. We don't sub-out much. Roughly only 30%. For my case, some of my works will even be 100% sub-out.
Q4: Why do you say 100% sub-out your works?
A4: Some of the sub-sub-cons, we are sure that they can handle and deliver the works. We simply pass the whole contract to them and then we make a percentage. These cases are clear cut.
Q5: So that means when you guys don't have the capacity to handle, you will sub-out to another sub-sub-con you know of?
A5: Yes.

3. **What types of work/trades do you typically sub-sub-contract out to others?**

Q6: What sort of trades do you sub-sub-contract out?
A6: My nature of work covers quite a lot of scope. I do a lot of BMS (Building management System), some electrical works and fire alarm installation and different types of trades, some dealing with wiring. We call it under building services.

L. Sui Pheng and S. Jia Zheng, *Construction Productivity in the Multilayer Subcontracting System*, Management in the Built Environment,
https://doi.org/10.1007/978-981-13-1831-3

4. **What guiding principles do you adopt to decide on (a) what, and (b) how much work to sub-sub-contract out to others?**

Q7: So for your case, any guiding principles to determine what to sub-sub-contract and how much work to sub-sub-contract?

A7: Well, not everything we are able to do such as the supply of equipment for the works, these are the things under the vendor and we also have our own vendor. For example, the fire alarm panels, we sub it out. These are under the equipment itself. The percentage against installation is also about equipment to installations is about 50–50.

Q8: So meaning that you do the wiring works, then for the installation of the equipment, the vendors will do it themselves?

A8: No. We purchase from the vendors in a set and we are unable to do it as there is a specification there. As such, we purchase from the vendors and then we do the installation. It is the same with the television set. They supply the television and then we do the installation, something similar like this.

Q9: So meaning that you sub-sub contract out the works when you don't have the expertise?

A9: Yes. Some are not our expertise. We are unable to do works such as programming as my men are not trained for such a work. These are under the specialist contractors that come in and do the programming of the equipment so that the equipment is functioning well and can then be handed over to the client. I would say it's a form of teamwork.

Q10: Hence, you purchase the equipment from the vendor and then you install it. And sometimes because of your firm does not have the necessary expertise to install it, then you will sub-out to the sub-sub-con who can do the job?

A10: Yes, that's right.

Q11: So something like this will occur, especially so for the case of the installation of the BMS system?

A11: Yes. This is a case where small installation we are able to do it but it is a waste of time and not being productive as my men are stuck at this job and cannot take on extra jobs. So cannot take on more work and will be stuck there doing such a job. So it is teamwork.

Q12: Any other guiding principles?

A12: The contract documents play a part that binds between the client and us, so this is an additional guiding principle as well.

Q13: Sorry Mr. L, what do you mean by this?

A13: Sometimes, the clauses are unfair to us but when you want to get the job, you really have to accept the clause that they have to define. This is because it is a business deal.

Q14: What do you mean by unfair clauses?

A14: Sometimes in the contract, they ask for the retention sum, like a down payment… and they hold this sum of money while you do the job until the end of the contract period. So this retention sum was hold on for quite a long time, even after the works have finished. So we have to follow the main-contractor's contract.

If the main contractors have any defects, we as their sub-con will also get penalised. This is why the main contractor's contract has always been unfair. Unless it's an NSC (Nominated Sub-contractor) where you directly tendered for the job instead of going to the main contractor and also depend on the client's situation.

Q15: So usually, you always enter into a contract with the main-cons?

A15: Some of our jobs we go through some main-cons while the others we direct tender for our works and thus we become an NSC.

Q16: So it is unfair in that sense that when the main contractor get the LD from the client, they will hold back your payment and won't release the progress payments to you? In fact, some of these main-cons will say that you cannot even get back your payment?

A16: Yes. This has always been the issue we sub-cons faced.

Q17: Is this quite a common problem that sub-cons faced?

A17: Yes, it is. So that's why your paperwork must be strong.

Q18: Paperwork as in… contractual documents?

A18: Yes. Sometimes, in the haste to get the contract, you overlook the contractual clauses and then you are in for a rude shock. By then it's too late as you are already bind by the main-con's contract. This is especially so when they give you such a contract quite last minute and this is quite normal. Hence, you must make sure that who you work for and whether that main-con is a reliable party. If not sure, better not take. Hence, it depends on your network. If you have a good network, it is easier to mitigate such potential pitfalls.

5. **In the midst of a tightening labour market and increases in foreign worker levies which has resulted in a squeeze in foreign worker numbers, do you sub-sub-contract more works out to others than ever before?**

Q19: Now under the midst of the tightening labour market like the reduction of MYEs etc., does your firm sub-sub-contract even more works out than before?

A19: Actually, like Mr. X (Main Contractor Partner), me and him join forces. We share our manpower together. This is one way we help each other to overcome this difficulty in order to survive. So we share our resources together. This is because the government is controlling the MYEs, levies are going up and it has almost doubled now. I hope that the government can do more to help us with this regard.

Q20: Besides combining manpower, does your firm subcontract even more works out to other sub-sub-cons?

A20: We do it as well.

Q21: So many subcontractors they are also doing it this way, sharing manpower?

A21: Yes. For the tactic of sharing of resources to be successful, both the main-con and sub-cons must trust each other and work hand in hand. So it's like a handshake, otherwise it will be very tough for us to survive.

6. **Would sub-sub-contracting out your firm's works to others affect your project construction productivity? How do you ensure your firm stays productive?**

Q22: Would sub-contracting your firm's work out affect your construction productivity? Any measures your firm has adopted to continuously stay productive?

A22: Actually, whether to sub-sub-contract really depends on the individual sub-cons capability. If the sub-cons can hand handle themselves, they won't sub-out. So sub-cons need manpower and their trust. Let me tell you that all the electrical installation workers are all Indians. Majority are Indians and the Malaysians have went back home as it is too far for them to go back home across the causeway. Hence, this boils down to manpower woes in the electrical installation works sector. Measures for us to stay productive are for us sub-cons to go for niche projects. For my case, I am more versatile as I am still doing builder works where me and Mr. X work together. If I get the entire work package, I will sub some of the works out to him or vice versa, a flexible arrangement. That's why I say that we share the manpower and works together. Hence, we can be productive and can also survive. Otherwise, it's tough luck for us. Equipment we also share.

Q23: So what about the works that you further sub-sub-out?

A23: We have to supervise. Actually at the end of the day, the key thing is payment. As long as payment is given timely, then it shouldn't be a problem already.

Q24: Alright, so allow me to recap. By sub-sub-contracting out your firm's works to others, you will actually improve your productivity?

A24: Should be able to say improve. But it will only cut down our overhead as we no need to employ so many workers and materials. There will be pros and cons. If the sub-sub-cons are no good and halfway through, they drop out, then you will be held responsible. Hence, it is good and bad, it depends on how you select your works and who you appoint as your sub-sub-cons.

Q25: Therefore, at the end of the day, it boils down to the issues of responsibility and reliability?

A25: Yes.

7. **Do you have a core team of sub-subcontractors whom you always work with? Why?**

Q26: Do you have a core team of sub-subcontractors whom you always employ, like what you have told me earlier, it could even be Mr. X, your main contractor partner?

A26: Yes. There is no fixed way as you, as the sub-con, can get the job and then you sub it out to your main contractors if both parties have agreed to work together. It can be vice versa.

Q27: So in our industry, is this a very common arrangement?

A27: Not very common. It depends on your discipline. That's why I say your contacts and discipline, where you dare to handle the job. Meaning that you dare to do the works. It depends on the nature of the work and the knowledge that you have in doing the works. If you don't have the knowledge, it's very difficult to do the job as you are more prone to screw-ups etc.

Q28: So you mean that it is necessary for you to have both the knowledge and experience on these works before you undertake these jobs?

A28: Yes.

8. **How do you help your sub-subcontractors to enhance their construction productivity, if any?**

Q29: How do you help your sub-subcontractors to level up in order to improve their productivity, if any?

A29: Well, because my subcontractors are not that experienced. They are almost purely labour, manpower based. So we sub-out to them and then they follow instructions for us. I just teach them and then get somebody to guide them all the way through. If you tell them to go training, they are reluctant to do so. They have to meet their daily needs. So now the government is enforcing that they have to go for those safety courses in order for them to work as these are the work-site safety issues. Without attending the course, they cannot continue working on-site. If it is not legislated and enforce, you think that these sub-cons and sub-sub-cons will go for training? No. They rather find more jobs to earn more. Nonetheless, it is good that the safety aspects they know. But nobody want to attend the course unless they have no choice but to go when they are forced by the government to go. I mean this is the government's technique to increase productivity but it doesn't suit us. It is the government's drive to do so as they see it as a form of long term investment to reduce workplace risks. Hence, they legislated and enforced it. Hence, the contractors have no choice but to comply. Even I myself as a licenced worker and you ask me to attend a course, I seriously feel that it is a waste of time. Even if the government put this on practice, it may not be suitable as they always keep changing their policies and their management as it is enforced.

Q30: Could it be that the sub-cons and the sub-sub-cons, their educational level is not that high. Hence, if you ask them to go for such courses, it may be futile as they may not even understand?

A30: No, I think it is simply a daily needs problem on earning more money to survive by doing more jobs. Try giving them $10,000 per day. You see whether they would like to go for these courses.

Q31: So meaning that the government must incentivise them to go for such courses in order to improve productivity in the industry?

A31: Yes, hopefully!

9. **How does your firm value-add to your main contractors whom you work with?**

Q32: How does your firm try to value-add to your main contractors, like to Mr. X's firm, whom you frequently work with?

A32: It's based on tender, how to add value? The lowest complying bid is the successful bidder…

Q33: So lots of firms now they adopt predatory pricing just so that they can get the job?

A33: Yes, they have to. No choice man! This year, due to the poor economy, lots of contractors will fold up their business. Their way of undercutting the tender price just so to get the job is not really sustainable in the long run. Foreign construction firms have more capital to tide through this crisis, unlike local firms are usually small firms where they have to join resources and pray hard that they can all tide over the economic crisis.

Q34: So I can say that as a sub-con, you don't really care much about value-add to your main-cons, but more on your progress payments?

A34: Yes, indeed. The sub-cons must meet the daily needs. If not, when the main-cons pass the job over to the sub-cons, who can the sub-cons survive?

Q35: Hence, sub-cons must take effort to get back your progress payment when it's due. If not, you really have to fight for your firm's survival or risk folding up your business?

A35: Yes, correct. So that's why I say the sub-cons are always at the losing end.

Q36: Have you ever try to upgrade to become a main-con?

A36: Well, it can be done and you have to work productively. However, you must be willing to take high-risks. If you want to sub your works together, do you have the time and resources? Whatever projects that can be done requires 2 key ingredients; Money and time. If you have the capital and manpower, no issue. If not, you cannot finish the project. Hence, you need people to invest in your firm in order to have sufficient capital. That's why sub-cons are always meeting their daily needs and always at the losing end.

10. **If your firm is a subcontractor, what and how would you want your main contractors to help you level up your construction productivity?**

Q37: As a subcontractor, how would you want your main-contractor to help improve your construction productivity?

A37: The main contractor will never help the sub-cons as they have their own "cost-centre". If you are out of their budget, will the main-con help you? Hence, it's all about time and money. The main-cons have their own difficulty. The main-con that you know about may also be a sub-con of others. The other factor besides time and money is independence—How you get the project?

Q38: So meaning that if you get the project by yourself, then obviously you earn more?

A38: No, it means that whether the project you got, will it help you make money? It's all about the money, the price. Now the market come out projects, but will anybody make money? Now because of the recession, all firms try to get their hands on a project, even if it's at a loss. Hopefully, they can tide over this tough period. I mean to be honest, subcontractors very difficult to do well. You just meet the daily needs; that is all. This is my point of view as a sub-con, where we finish our jobs and don't get our payments etc. As long as you continue to sustain or maintain and wait for the right opportunity to shine. Nobody will help the sub-cons. It's all about your luck as well.

Q39: So just to recap question 10, as a sub-con, it really depends as the main-cons themselves will also face difficulties and so it all depends on the sub-cons themselves to pull through, am I right?

A39: Yes, and also the sub-cons must be knowledgeable on what they are doing. If your employee not knowledgeable, you will also face the same issue by not being competent in your job, then you lose your competitiveness and then you will die-off. There also a lot of factors at play as well.

Additional Points

Q40: In general, what's your opinion on the issue of productivity in our industry currently?

A40: Construction all rely on manpower, where got productive? Without the manpower, how to do the job? If the tender price not good, how to make money? Then how would you afford to send your workers for such government endorsed courses to improve productivity? Like for example, SMRT, they want to control the budget and thus they don't spend money on maintenance. Now, they faced all sorts of maintenance issues. My point is that if you want to control an aspect, it will have an effect or consequence downstream. Similarly, the government because they want to control the number of foreign workers idling in our country, they raise the levy. I think it could be due to after the completion of the IRs, too many workers left idling around, so they want to send them back ASAP. This increase in levy and MYEs caused a sharp reduction of foreign workers. Now, it has caused a manpower shortage in the construction sector.

Q41: With regards to the subcontracting practices in Singapore, what are your views on this matter?

A41: You want to make money now it is quite difficult, all due to increase in levies, lower MYEs increase in CPF contributions and also then recession etc. Everything is going up but not the tender price. Do you think you are able to sustain? No! The tenders all during recession are all lowly priced. Only the foreign companies are capable of pricing this low. Our sub-cons have to follow the main-cons and we will be sandwiched. Let's say our main-cons get the tender successfully but the jobs cope so tough where the scope is wide and the price is low. So you need more manpower to ensure optimal operations, just like MRT projects. In MRT projects, their specifications are very stringent. This leads to a few main-cons went bust in doing such MRT projects as they lose lots of money. So the sub-cons that follow these main-cons will suffer badly. Hence, these sub-cons are eventually not paid. Sub-cons have limited resources such as limited capital. Therefore, sub-cons have a hard time sustaining themselves.

End of Verbatim Report 12

Annex 17: Interview Transcript 13 (Main Contractor I13)

Date of interview: Wednesday 3rd Feb 2016.

Position of Interviewee: Senior Contracts Manager.

1. **Does your firm sub-contract works out to others?**

Q1: Does your firm sub-contract works out to other subcontractors?

A1: Yes.

2. **What percentage of work does your firm typically sub-contract out to others?**

Q2: May I know what is the percentage of work that your firm typically sub-contract out to your subcontractors?

A2: As a main contractor, we sub-contract out all the works except for indirect temporary works such as housekeeping and safety works which may be carried out by our direct workers under our site management's control. Therefore, 100% of the works are sub-contracted out except for management and the aforesaid indirect temporary works.

Q3: So works such as general maintenance of construction site, only then will your company have their own general workers. If not, the rest of the works will be subcontracted out?

A3: Yes, that's how we do it.

3. **What types of work/trades do you typically sub-contract out to others?**

Q4: So I presume that you sub-out all sorts of trade since it's 100% subcontracted out?

A4: All trades are sub-contracted out as we are the main contractor involved mainly in project management.

Q5: Hence, when you sub-contract out 100% of your works, you are simply doing project management? Am I correct?

A5: We are more of doing project management style of main contractor, where we coordinate various trades together, put various shop drawings together, then get

L. Sui Pheng and S. Jia Zheng, *Construction Productivity in the Multilayer Subcontracting System*, Management in the Built Environment,
https://doi.org/10.1007/978-981-13-1831-3

the work flow organised, prepare the overall programme. Hence, by doing this, every subcontractor's work, they focus within their own timeframe. So by doing this, we allow our subcontractors to concentrate on their works. Of course, their workers will be specialised in their trades right? So usually, if it is the main contractors, they will handle many different types of trades in their projects. For us, we feel that it may not be as efficient and productive, as compared to subcontractors having their workers concentrating on their speciality. So this is on the positive side. However, if you compare profitability, there is a possibility that by having our own direct workers such as structural, concreter, bricklayer, tiler, all under our control. If productivity we want to maintain at a very high level, we must ensure that our own supervisors and foreman are all up to speed and they are very capable to push the works. So productivity doesn't rely on the workforce itself, it relies a lot on the management. If it is the subcontractor's management, they will definitely want to do things as fast as possible, do it the first time so that their cost can be as low as possible so that they can earn as much money as possible. That's where there is pros and cons of subcontracting.

Q6: So that's where your side try to strike a delicate balance between ensuring maximised efficiency in getting the job done within the time frame and also ensuring that the subcontractors don't take short-cuts. Am I correct?

A6: Yes. So that's why when we do our site management, we will have supervisors and engineers trained to or delegated to control the quality of works and carrying out works within a certain timeframe.

4. **What guiding principles do you adopt to decide on (a) what, and (b) how much work to sub-contract out to others?**

Q7: So for your case, any guiding principles to determine what to subcontract and how much work to subcontract?

A7: There are 2 main criteria.

A. Works are packaged out by trades and level of work difficulty
B. Safety, quality, price, sub-con's current commitment and company's financial health. In addition, familiarity with the sub-contractor can help to ascertain the suitability of the sub-contractor being able to carry out the works.

Background checks also need to be done to be sure of the subcontractors' financial status; we compile and progressively update our approved vendor list and make periodical review; make a performance assessment upon the completion of each project.

Therefore, how much work to give depends on the trades and work packages that have been planned by the site team. Certain trades will go to certain specialist subcontractor. Certain trade may be given to 2 or more subcontractors depending on the scope of works that were given.

Hence, we subcontract everything. So how we package them is that we will normally package them by trades such as wet works like brick layer and tiler etc. can be under the same party etc. Then the dry works can come into another package as well. So generally the wet works subcontractors would have lots of specialist workers who are just skilled in laying bricks and tiles, doing screeding and stonework etc. So we need to package out works very reasonably and accurately in order to prevent any overlapping of works and specific trades. For example, I don't give ceiling subcontractors whom we call dry works as ceiling and partitions can be with the same guy. But wet works and the ceiling guys may not do the same together. So we must make sure that there is no clash in the trades itself. We try to prevent such clash itself. Even if we don't try to prevent this clash, it's still workable but you must ensure that the subcontractors must be able to procure their own workers and manage the quality and work within our timeframe. So generally the packages that we award are already designed to give specialised areas. Concreters, formwork and rebar guys, we will definitely award to one company for example. And within that, they themselves will have their own bar bender, concreter and formwork guys. So likewise, wet work we also try to pick up wet works in another group.

Q8: So that means you rather sub-contract out trade-based works for example subcontractors that specifically specialised in wet works, wet trades and so we sub-contract the entire package to them? Similarly for dry trades as well?

A8: Yes. So generally in a project, wet works are probably taken care by probably one or 2 subcontractors. Likewise, partition and ceiling works (also known as dry works) will also be taken care of by another group. Then you also have specialist works such as acoustic glass or fire glass, fire-rated ceiling, it can be done by other specialist contractors as well. Then you have another category of works such as cladding works, glazing works. Hence such works will be required in another group. So for example, glazing works or curtain wall works, we also need to consider that certain subcontractors may not be able to carry out certain level of work difficulty. So you may have certain difficult design, you may have to depend on more reliable and professional subcontractors who can better manage, to cope with certain difficult detailing. Hence, we cannot choose sub-cons blindly, we have to look at their expertise. So another additional guiding principle is expertise.

Q9: So besides the expertise, the work package by trade and price, any other guiding principles?

A9: You have to look at the company's financial health in order for us to determine how much work to sub-contract out and rely on subcontractors. We are not going to award them big value of works to these sub-cons if they are not able to cope, by sub-dividing further into different areas for different people. Unless if that particular sub-contractor is extremely capable and we have more confidence, we will probably give them the full package. Hence, their financial background is important to us, their quality of works is important to us, their timely delivery is important for us. Their safety ability is important for us. Their concern of the environment is also important for us. We have these 5 guiding principles that we always adhere to; quality, cost, delivery, safety and environment. Hence we try to balance all these out.

Q10: Why don't your firm retain some of the works in-house, why 100% sub-contract out?

A10: That's because our style of works and our background, if we were to do it that approach like some other local main contractors like X or Y, they have more direct labour, like wet works and concrete works and certain finishing works as well other than specialised works, they try to carry out the works on their own. I also replied that if you were to have a direct workforce, there is a long term turnover of projects, we must ensure that your workers are always fully utilised. If they are not fully utilised, for main contractors it would not be financially viable. This is one consideration. The second consideration you have to look at your own in-house and supervisory capability. If you have your own direct workers, for your site to be very productive, you must make sure that the site supervisors and the in-house foreman are very well trained and competent. This is another key point. Meaning that the workers can finish the works within a certain time. So these are things that balanced out, whether you can deploy workers and manage workers on your own. So for companies of our size, we normally don't.

Q11: This is because your firm has multiple huge projects of very large value, so it doesn't make financial sense to own the direct workers?

A11: Yes. So it's basically a matter of being able to fully utilise all the workers at any point in time. If you continuously have numerous jobs, you are able to own and maintain these workers. That will be a good point where you have competent foreman and supervisors to ensure that the works are being carried out at fast pace, good quality, safely. So if all these points can be achieved, then we can have our own direct workers.

Q12: So for now, your firm's Management feels that they are unable to balance some aspects of it?

A12: Yes, in terms of supervision and ensuring that the workers are fully utilise all year round.

Q13: That boils down to the uncertainties faced by contractors on being able to successfully tender for new jobs as well right? Especially so when there is a recession going on and the number of jobs in the market is shrinking?

A13: Yes. We can feel that as well. Not much jobs running around. We have not gotten jobs for the past one year.

Q14: So your firm is now clearing the existing backlog of jobs it has currently?

A14: No. We still have a few big jobs ongoing but we don't have new jobs coming in yet. And this is the sign of a looming recession.

Oh yes, one more point on how we package for specialist trades, we cannot award just because it is dry trade such as dry ceiling where the ceiling works are fire –rated ceiling brands where normal subcontractors are unable to do. They may have light sensors where normal sub-cons may not have. You need to consider this when you package for works to be subcontracted out.

Q15: So this means that for such specialist work like fire-rated ceilings, despite having subcontracted ceiling works to a subcontractor already, you have to further sub-contract out to the specialist contractors?

A15: Yes, correct.

Q16: This also means that the specialist subcontractor will also have to liaise with the packaged subcontractor who has got the ceiling job package as well to do up the ceiling?

A16: Yes.

5. **In the midst of a tightening labour market and increases in foreign worker levies which has resulted in a squeeze in foreign worker numbers, do you sub-contract more works out to others than ever before?**

Q17: Now under the midst of the tightening labour market like the reduction of MYEs etc., does your firm sub-contract even more works out than before?

A17: The restrictive conditions may affect the subcontractor's pricing and for that reason we may find additional subcontractors or plan for more work packages to find cheaper alternatives. In any case, other than housekeeping workers, we sub-contract all work packages. I also think that this question is more suited for companies that have both direct workers carrying out the permanent works and subcontracting may be a bit more suitable to answer your question. For us, as you know already, we subcontract all our work packages. So yes in general, tightening labour markets will affect all contractors. Even our subcontractors are affected as they are not able to cope with the current work demand they have. Part of your question may be related to our case on shortages of workers… If you know that particular works, because of this tightening labour market situation, instead of awarding to one particular contractor, we may have to rely on 2 subcontractors. We may actually have to divide a work package into several packages.

Q18: So you are actually referring to the capacity of the subcontractors?

A18: Sometimes because the subcontractors are unable to procure the optimum number of workers for their own workforce. So previously they had say they can provide 100, but now limited to 50 workers and you have a certain dateline to meet. So considering you have a deadline to meet for your work package, you have to consider how much work can be done by these 50 workers in this period of time. If you know cannot, then you have to tell the subcontractors that we are unable to give you the entire work package, we will have to sub-divide to 2–3 subcontractors. That will help us reduce the risk of subcontracting failure such as non-performance, where if one of them is in breach of the contract, at least I have the other 2 to take over the works. We balance out the risk.

Q19: So this is a form of risk minimisation?

A19: Yes.

Q20: Just to confirm, earlier in the interview, you mention that your firm has direct workers. Are these direct workers also known as general workers?

A20: Yes. But these are small in numbers. Say only about 50 workers where they will be distributed to various sites. They are in charge of safety, site access, clearing of rubbish etc.

6. **Would sub-contracting out your firm's works to others affect your project construction productivity?**

Q21: Would sub-contracting your firm's work affect your construction productivity?

A21: Yes, we believe that it would strongly boost productivity. Firstly, if the various trades are carried out by trade skilled specialists and specialist sub-contractors who are more familiar with nature of the trade and thus, they will spend the least time to carry out the works. If such trade works are not sub-out, there is a tendency for failure in the event that the site management is unable to coordinate well. This is especially so if the workers are deployed directly to carry out the works. We also feel that if you consider these skilful workers that are managed by your subcontractors, it will definitely be beneficial in terms of time and quality as they are more familiar with their works. Then let's say if it's in the negative side, if we want to keep such work in-house, we must ensure that we have properly trained site supervisors and management. If they are unable to coordinate well, they will definitely fail.

Q22: So what do you mean by site management?

A22: Right here, this is our HQ. Our company will actually have project team and they will be stationed at project sites. So within the project site, there will be supervisory staff, engineering staff, drawing coordination staff, PM, PD, QS and QC specialist and so on etc. So among these group, let's say there are direct workers, we must ensure that we will good foreman and supervisory staff to ensure certain trait of works that come under our own scope is carried out very efficiently. So in our case, we normally manage our subcontractors directly, instead of managing our won workers. So for certain companies, where they have their own in-house supervisors, in-house foreman, they must first make sure that the in-house foreman knows how to get things done smoothly. That is the key to success.

Q23: So mainly your firm just liaise with the management of the sub-cons?

A23: Yes

Q24: so then it's actually the sub-cons' supervisor and foreman that get things done?

A24: Yes. But in all, having subcontractors in the positive side is that they will enable higher productivity as they are specialised in a particular works and usually they want to get things done fastest, shortest and easiest and cheapest way so that they can get as much profit as possible.

7. **Do they have a core team of subcontractors whom they always work with? Why?**

Q25: Do you have a core team of subcontractors whom you always employ?

A25: Yes. Their previous experience working with us would enhance their understanding of our general objectives of quality, cost, safety, delivery and environment. Regular subcontractors would also be more open to cost negotiations, assist with necessary acceleration measures to support the project as opposed to new partners. Most importantly is familiarity and their understanding of our objectives and their understanding of our requirements as well. So for example, sub-con A has been working for us for years, they know our in-house safety and regulations very well, they know our quality standards and our requirements and they will understand how time is always important for us and they understand the importance of coordination. So a few companies of various trades will tend to follow us, so there is a tendency to work with these same groups.

Q26: So what happens if your core team of subcontractors are also faced with full work capacity but then there are still jobs that needs to be done. How would your firm solve this?

A26: Based on current conditions, it won't happen as recent times, market is quite flat, so we won't experience this. However, in 2007–2008 was the super-hot period where busy period or not, our regular sub-cons tend to be there for us. They will be able to manage by getting outsourced workers on their own and still support us.

Q27: Nonetheless, in the event that they are not available and you need to look for other sub-cons, what are your requirements in selecting your desired sub-con?

A27: We will definitely check on their financial background, their project experiences and their track record, these 2 key points and in particularly, the kind of projects that they have carried out before and if it is similar in nature to our current project. For example, if you know that they are factory subcontractors, they always carried out factory works where it is not super A quality works whereas we have requirements to construct Grade A office building. Hence, we must ensure that the sub-cons have such experiences before and they are not a $2 company, so these are all sorts of considerations, interview them and ensure that their financial backups and ash flow is there before we commit to any sub-cons that are new. Particularly, we will also look at their BCA registration, what is their financial standing etc.

Q28: So what is the desired amount of paid-up capital that you are looking for when selecting a new sub-con?

A28: It depends on the value of the work package that we are undertaking. Our wet-work package if they are within $1 million, of course minimum paid-up capital is between $200,000 and $500,000.

Q29: So the paid-up capital is a means to find out how serious the sub-cons are in this line of work, am I right?

A29: Correct. Another point we will look out for is the sub-con's ability to procure performance bonds. This means that they have enough money in the bank. Only when they have enough money in the bank, will then the bank give a bond to them. So bond value is equivalent to money in the bank. So that is one more security that we look for as to whether they are financially stable. Sometimes, when you are in delay, when you call your core team of sub-cons who have good work experiences with you, they tend to support you when you are in trouble especially if there is an urgent need to complete the works within say 10 days, they can bring 50 workers to you, work at night and thus they tend to support because the relationship is there.

8. **How do the main contractors help their subcontractors level up to enhance productivity, if any?**

Q30: How do you help your subcontractors whom you always subcontract works to, assist them to level up in order to enhance productivity, if any?

A30: Site coordination and detailed coordinated shop drawings would enhance construction productivity and reduce re-works. Daily coordination meetings, timely interim payments to ensure the subcontractors' workers are timely paid and kept motivated, provide safety briefings prior to work commencement to avert work stoppages that are caused by safety non-compliances and incidents. In addition, providing information to subcontractors on various training programmes for their workers to upgrade, considerations to deploy mechanical equipment such as boom lifts / scissors lift instead of scaffolding and create good access and safe scaffolding for the subcontractors' usage. As mentioned, one of the key factor is the coordinated shop drawings, where it is one of the important factors in terms of getting work done properly and accurately on site with proper quality. So that will actually increase the construction productivity. Ensuring daily coordination and things like that and when, where and what areas to work, how to go around from one room to another, pushing from one trade to another.

Q31: At the end of the day, it's all about the communication?

A31: Yes. It's also about the money, via timely interim payments, so that the sub-cons and their workers are always well paid. We also have daily safety brings prior to commencement of works, so otherwise when there is no proper safety briefings and reminders, there tend to have a lot of work stoppages due to non-compliances and incidences. So these are the key points that we tend to highlight so that the work stoppages do not happen.

Q32: So speaking of payment, it's a very thorny issue in this industry. There are instances where main contractors hold back their sub-con's payment in full even after the work is done. How does your firm deal with payments?

A32: One of the reasons sub-cons tend to follow big companies like us is due to prompt payments and reasonable pay masters. We don't usually screw payments for them unless they over claim, failed in the works and that's when we may end up in disputes and arbitration. Otherwise, if you're a regular, usually your payments will be quite prompt.

Q33: So it's the assurance and the credibility that motivates the sub-cons to do well?

A33: Yes. So that's the reason that they are able to probably give us a better discount and pricing as compared to working with the other main-cons. How you enhance the construction productivity is to deploy more mechanical equipment. For example, ceiling works, we tend to use scaffolding which we spend time erecting and dismantling it vs. the scissors lift that just goes up and done.

Q34: But then again, there's a limit to how much mechanisation that can be done right? Because some of the trades the mechanisation is just not available. Would this be an issue to you?

A34: I mean when mechanisation is fully available, you try to fully utilise it. But that doesn't mean that in an open big ceiling, just by putting machinery, it will enhance productivity. We would have to calculate the overall productivity even when we use machines vs. if we use a full board bridge style scaffolding. You are able to carry out many works on the ceiling in one go, if not machines also go up and down and it will also be too crowded and thus you can't move. Hence, mechanisation doesn't automatically equate to high productivity.

Q35: So you must depend on what sort of works you are doing?

A35: Correct. Do a rough productivity study for all the works involve in a particular area before we commit to what sort of methodology that we are going to carry out.

9. **Is your firm also a subcontractor?**

(Not applicable, this firm is a main contractor.)

10. **If your firm is also a subcontractor, what and how would you want your main contractors to help you level up your construction productivity?**

Q36: So has your core team of sub-contractors gave feedback to you, the main-con, on how you can help them to improve productivity? Do they even take this initiative to do so?

A36: Not applicable.

Additional Points

Q37: So as a senior contracts manager, you are always involved in all these sub-contracting practices, deals etc. In your opinion, what do you think is plaguing the subcontracting practices in our industry?

A37: I think generally all over the world, subcontracting is very predominant and it will always be there. Mainly it is because of niche market for specialisation works. Like what I told you for building industry where multiple trades are involved, it is very difficult for the main-con to actually manage all the various trades and house these specialist workers within their own portfolio because of the continuity of business is not there. We may have one project for 1–2 years and then suddenly, you realise that these workers have no more jobs to do just because you have no more works to do. Subcontracting is always there because of specialisation.

Q38: What about multiple layers of subcontracting that is prevalent in our industry, does it affect our productivity?

A38: I don't think it affects productivity but more of cost. This is because of various value-add that you go up the channels so that everybody is making certain level of profits. So you will definitely feel that the price will actually be higher at various stages. So sub-sub and then subcontract the works out, so eventually it will come back to the main-con and then to the owner. So definitely things will be more expensive right? And so things will not be as efficiently priced.

Q39: So what you are saying is that it is not efficiently priced because of the many layers of mark-up?

A39: Yes. But not necessary sub-cons rely on their own direct workers as well. Sometimes they also have to package out things to their sub-sub and that's because the number of skilled workers each sub-con can have on their own. So somehow even if they package downwards, they should be able to manage within their cost as they would have priced for such possible costs of labour, material etc. So generally, the mark-ups will be very slight.

Q40: So the main-cons have already considered all these various mark-ups in various stages?

A40: Yes. But we will only consider what our sub-con's price is and they will have to work within the budget as well.

Q41: So this many tiers of subcontracting, so it has caused the price to become inefficient? But it does not affect construction productivity?

A41: Yes. Generally the many layers of subcontracting should not be affecting construction productivity if your coordination, your shop drawings, the worker's level of skill and the use of machinery are there. In fact, It should be better.

Q42: All because they are doing their own specialised trade?

A42: Correct.

Q43: What are your views on construction productivity in Singapore?

A43: I think there's still a lot of room for better productivity. One of the main reasons is we are relying workers from 3rd world countries. If you compare productivity in the developed nations like Japan, US and Australia, somehow the skill levels are much higher and in terms of workers' respectability, they are different. Their workers are treated differently. They are sort of a profession. You would be surprised that certain developed countries, the workers can earn up to $6000 monthly compared to what the worker scan earn in a 3rd world place or for the foreign workers in Singapore. So one of the major reasons is because how much remuneration you are giving to the core group of workers. The origin or source of the workers are definitely different. So I think there's much more room to improve if the right type of workers come on board.

Q44: What do you mean by the "right" type of workers?

A44: It's about having the skilful workers but somehow where work is carried out in developed countries, maybe it is better. Somehow works are carried out more efficiently in developed countries.

Q45: Could it be due to the education level of workers, the mind-set are different?

A45: Yes, also the training given is different, motivation is different. I think there's way much room for improvement.

Q46: So if we aspire towards having the "right" type of workers like those in developed nations, won't it push-up the prices of construction works?

A46: Definitely, in developed countries, construction cost is way more expensive than in Singapore. On the other hand, land cost is cheaper in other countries but not in Singapore due to land scarcity. So things will still balance out. This is because labour cost in developed countries are much more expensive but cheap land cost whereas in Singapore, land cost is expensive but labour cost is much lower than other developed cities. Not much change in material cost as well.

End of Verbatim Report 13

Annex 18: Interview Transcript 14 (Subcontractor I14)

Date of interview: Thursday 4th Feb 2016.

Position of interviewee: Manager.

1. **Does your firm sub-sub-contract works out to others?**

Q1: Does your firm sub-sub-contract works out to other subcontractors?
A1: Yes, we do.
Q2: So what is the nature of your business?
A2: I do fire protection systems.

2. **What percentage of work does your firm typically sub-sub-contract out to others?**

Q3: May I know what is the percentage of work that your firm typically sub-contract out to your subcontractors?

A3: Usually around 30%. But it depends on our capacity. If really cannot take on more jobs but still have jobs for us, we will 100% sub-out to reliable sub-sub-cons that we know of and have worked closely many times with them together before. Our labour cost will be 100% and our material cost will be 10%.

3. **What types of work/trades do you typically sub-sub-contract out to others?**

Q4: What sort of trades do you sub-sub-contract out?
A4: Usually it's the fire protection works.

4. **What guiding principles do you adopt to decide on (a) what, and (b) how much work to sub-sub-contract out to others?**

Q5: So for your case, any guiding principles to determine what to sub-sub-contract and how much work to sub-sub-contract?

A5: The main principle is pricing. Besides that, we take into consideration their manpower availability and capacity to do the job and also the skill to do the job.

L. Sui Pheng and S. Jia Zheng, *Construction Productivity in the Multilayer Subcontracting System*, Management in the Built Environment,
https://doi.org/10.1007/978-981-13-1831-3

5. **In the midst of a tightening labour market and increases in foreign worker levies which has resulted in a squeeze in foreign worker numbers, do you sub-sub-contract more works out to others than ever before?**

Q6: Now under the midst of the tightening labour market like the reduction of MYEs etc., does your firm sub-sub-contract even more works out than before?

A6: We try to do the installation works by ourselves and if we unable to handle the workload or got specialised components like fire alarm, we will still sub-out to the specialist contractors.

Q7: I thought fire alarm is a specialised component by itself?

A7: No, some fire alarms work hand in hand with the fire protection systems.

Q8: Then under such difficult circumstances, do you further sub-sub-contract your works to other parties?

A8: We will not sub-out because the levy is expensive and then the tender price now very bad, all making a loss. Now most subcontractors also do their work in-house, can cut cost and cut other miscellaneous stuff. Even if you sub-out, got pros and cons. Pros is that you no need to worry about your workers' levy as it is not your problem anymore, however you incur a higher cost as you need to price in overheads for the sub-sub-con whom you are subcontracting to.

6. **Would sub-sub-contracting out your firm's works to others affect your project construction productivity? How do you ensure your firm stays productive?**

Q9: Would sub-contracting your firm's work out affect your construction productivity? Any measures your firm has adopted to continuously stay productive?

A9: We will try to remain as productive as possible. Usually it will be more productive as the sub-sub-cons, who are usually the specialist contractors, have the necessary expertise in doing the job and can finish the job within the designated time frame or when the job needs to be rushed for completion. Hence, it results in increased efficiency. However, there are times when the sub-sub-cons have too much workload, then sub-out to them also not very efficient as they cannot do finish the job as soon as possible. In addition, we also closely monitor our sub-sub-cons and also solicit feedback from them to remedy any potential issues that they faced in the course of their work to minimise hiccups that can delay the works. If the sub-sub-cons are uncooperative, then we will feedback to their bosses and then we will not give any more jobs to them in future.

7. **Do you have a core team of sub-subcontractors whom you always work with? Why?**

Q10: Do you have a core team of sub-subcontractors whom you always sub-out to?

A10: Yes. They are our regular subcontractors. We are familiar with them as compared to other sub-sub-cons whom we never know before. It's more on reliability.

Q11: If your core team of sub-sub-cons do not have the capacity to take up more jobs, but you still have jobs coming in, how would you do?

A11: I will ask them to try and accommodate my requests. If really cannot, then I have no choice but to sub-out.

Q12: Then how would you evaluate the various sub-sub-cons to ensure that you get the right one?

A12: Most importantly is the price, then ask about their reputation in the industry and then if it satisfies your requirements, then you sub-sub-out to them. Actually, in the construction line, it is simply "trial-and-error". This is because sometimes, the intended sub-sub-con that you hired cannot really perform the tasks or worse still, the intended sub-sub-con further sub-out your job to another party, which is very hard to detect until it's too late to do anything. This is prevalently used in Singapore.

8. **How do you help your sub-subcontractors to enhance their construction productivity, if any?**

Q13: How do you help your sub-subcontractors to level up in order to improve their productivity, if any?

A13: We usually conduct close monitoring and supervision of their works. In addition, we also update our records with the latest information if there are any changes.

9. **How does your firm value-add to your main contractors whom you work with?**

Q14: How does your firm try to value-add to your main contractors, like to Mr. Y's firm (Main Contractor Partner), whom you frequently work with?

A14: We work together closely where we have weekly meetings to update site progress and raise any issues that needs to be addressed.

10. **If your firm is a subcontractor, what and how would you want your main contractors to help you level up your construction productivity?**

Q15: As a subcontractor, how would you want your main-contractor to help improve your construction productivity?

A15: The main contractor can plan and schedule their work sequence properly and conduct adequate briefings to ensure that we are well-informed at all times in order to minimise any miscommunication that can cause costly errors in our works. There are many main-cons that are like this.

Additional Points

Q16: With regards to the subcontracting practices in Singapore, what are your views on this matter?

A16: Nowadays, subcontracting is very common. However, I think efficiency might be lower because you are actually informing others on how to do the job and then this piece of instruction is passed down to any layers of subcontractors, sub-sub-cons etc. Hence, such instructions are easily miscommunicated or at the very worst case, lost in transition. Let me tell you, the construction line is a very messy line. At the end of the day, it all boils down to payment issues. If you don't have the timely progress payments, very hard for your firm to survive in the long run. When you become a contractor, you must have the necessary financial capital to do so.

Q17: What are your views regarding construction productivity in Singapore?

A17: You must see what trade you are looking at. Sometimes, like got the productivity score as mandated by BCA, best to score as high as possible. Actually, with regards to productivity, it's actually differs from trade to trade. The government talks in a generic sense on productivity. However, there are so many different works/trades in construction. Hence, different trades will have different ways of doing it. Hence, it is very difficult to talk about productivity as a whole, as each individual trade has its own productivity. For example, last time electric drill got wire and socket, always need to connect to the plug. Now we simply sue battery-operated drill. More portable and flexible to use anywhere, hence higher productivity. Another case would be tiling. Now tiling can only be done manually right, so hard to be productive. No doubt it is good to be productive, but it can also rob our jobs. Take for example the PPVC bathrooms, now just install in like lego as it is precast. Thus there is no need for plumbing and sanitary fittings and installations anymore, where these 2 are trades that I currently depend on for my livelihood. If the PPVC components are installed by us, then at least we are still able to survive. There are some trades that benefit from these pre-cast components, such as the tower crane operators, they will be able to earn a lot based on the presence of such precast components that needs to be hoisted on-site. Last time, contractors rented 40 tonne cranes, now they rent 200 tonne cranes. In conclusion, each trade is different from each other.

Q18: Just now you have mentioned that when work gets subcontracted out, the tender price is very bad in today's market that is undergoing a recession. If so, will it affect the progress payments of the subcontractors as well?

A18: It all depends on the main-con. Some main-cons are bastards. They will give unreasonable excuses like because they are not being paid by the owner, you as their sub-con, will not get paid as well. The better main-cons will follow the SOP Act and pay the sub-cons after 35 days.

Q19: So if the main-cons failed to give you progress payments, will you activate the SOP Act against the main-cons?

A19: Actually small fry sub-cons like us are at the losing end. It depends on the financial status of your main-con. If those main-cons are in the red and on the brink of bankruptcy, then you have no choice but to employ SOP Act to get back the payment from them.

End of Verbatim Report 14

Annex 19: Interview Transcript 15 (Subcontractor I15)

Date of interview: Saturday 20th Feb 2016.
Position of Interviewee: Managing Director.

1. **Does your firm sub-sub-contract works out to others?**

Q1: Does your firm sub-sub-contract works out to other subcontractors?
A1: Yes, we do. This is because not many firms can keep so many workers.

2. **What percentage of work does your firm typically sub-sub-contract out to others?**

Q2: May I know what is the percentage of work that your firm typically sub-contract out to your subcontractors?

A2: Some of the work we fabricate and we sub-out the installation works to our sub-subcontractor. The percentage would be around 20–40%.

Q3: What is the nature of the works that your company undertakes?

A3: We are a subcontractor. For our company, we are mainly dealing with aluminium, glass and stainless steel works for the building. Different companies will do different works like ceiling, doors and other equipment. Some would also be doing carpentry work while our neighbour is an M&E contractor. So we deal with louvres, window, glass, panelling all these works. Meaning that we do exterior building facade, the inside like railing, etc.

3. **What types of work/trades do you typically sub-sub-contract out to others?**

Q4: What sort of trades do you sub-sub-contract out?

A4: Usually we will sub-out the installation works as mentioned earlier. Sometimes, we sub-out fabrication work if our factory cannot cope. This is because you always spilt into production and site team. If your production team capacity is full, then your production side you must sub it out to other subcontractors. Then your installation team, if your site team is too busy and more things to do, then you have to sub-out as well. This is for construction. For factory wise, I think those

L. Sui Pheng and S. Jia Zheng, *Construction Productivity in the Multilayer Subcontracting System*, Management in the Built Environment,
https://doi.org/10.1007/978-981-13-1831-3

standard production, they don't really have a site team. It's just like a factory where they just produce only, they don't really involve in the site work.

Q5: Like those work activities that is ongoing downstairs, the various metal works?

A5: Yes, that is under factory production.

4. **What guiding principles do you adopt to decide on (a) what, and (b) how much work to sub-sub-contract out to others?**

Q6: So for your case, any guiding principles to determine what to sub-sub-contract and how much work to sub-sub-contract?

A6: It depends on our workload and it also depends on the amount of jobs that we have. If we currently don't have other jobs, we don't sub out. Let's say during a particular month, so many projects, of course we need to sub-out. Also, it is very subjective to say how much to sub-out as it will depend on our business, our strength and of course not enough must sub-out. If you have workers but you have no job, your worker must do work right? So we first leave it to our own source to do the job. If we cannot cope, then we sub-out.

Q7: From the word "source", you meant in-house workers?

A7: Yes. For us, we have 40–50 workers, thus we will have to use our own workforce to do the work. If we not cannot and not enough workers, then we sub-out. All subcontractors the same. We cannot let our workers do nothing then you sub-out, then you still have to pay them salary. So it really depends.

Q8: So I would say that it depends on the capacity of the firm?

A8: Yes.

5. **In the midst of a tightening labour market and increases in foreign worker levies which has resulted in a squeeze in foreign worker numbers, do you sub-sub-contract more works out to others than ever before?**

Q9: Now under the midst of the tightening labour market like the reduction of MYEs etc., does your firm sub-sub-contract even more works out than before?

A9: It also depends. We need to get the MYEs from the main contractor. If don't have, like now getting lesser and lesser and the works are increasing, then we sub it out. It is also very subjective as if we want to keep so many workers, then now got recession and some small construction firms start to close down. The reason is lesser building jobs than before, less residential jobs. Then those small firms that cannot survive they have to close shop.

Q10: So in other words, your firm sub-out the works even though there is a reduction in the numbers of foreign workers available or it still remains the same as before?

A10: In fact, it is harder to get workers. Those firms that have workers will demand higher salary. Last time for example, we sub-out, they demand $70, now they demand $100. So indirectly, it incurs more to sub-out, it increases our cost to sub-out. So it's a supply and demand issue. Singaporeans don't do the work. Zero. All foreign workers. So when you tighten the influx of foreign worker, everybody will bring up their price, because it's about supply and demand. The sub-con will demand higher price. So end of the day, all these costs will be transferred to the consumers. In fact, our cost has increased when they want to tighten. Basically, a lot of issue, not only due to the tightening, but also due to the regulation.

Q11: What regulation are you referring to?

A11: It's the MOM regulation. The MYE and safety requirements will incur all the cost. I tell you for instance, if somebody want to change the lighting bulb, you need the risk assessment and safety supervisor to change one bulb. Then the fellow who bring the ladder to climb, also need somebody to hold the ladder, that is the regulation. Then the guy that climb also need a safety belt under the supervision of the safety supervisor who just stand there and watch. So the 3 person cost you see how heavy, is it productive? In Europe and Australia, only 1 person do the work. They take care of themselves and they are the safety supervisor and everything they do themselves. So in fact, the 3 workers' salary add up together is equal to or even higher than one Caucasian worker. So it is not productive at all. And you know that our foreign worker, the productivity is very low. It's not like the Caucasian where they have the passion and concentration to do their work.

Q12: So our foreign workers they come from developing countries, they don't have the awareness and passion to get work done properly?

A12: Yes. They actually are farmers and they come here to become workers. You ask them to do construction trades, they don't know. Hence, you must train them. Then when you finish training them, they reach their maximum length of stay here and they have to go back already. Hence, we like a training centre for them. All these issues make us very hard.

Q13: So I would say that it's the rigidity of all these regulations that are causing these problems?

A13: Yes, the authorities and the government must go down to the ground in order for them to understand the problem. The government must have their own construction company, only then they will encounter all the problem and even the sub-contractor company as well. Then they will know what the actual problem is.

Q14: So I would say it is the regulation that is not feasible which is causing these problems?

A14: Not really due to the inflexible regulation. If a lot of Singaporeans, like the Chinese, love "face". Whereas the Caucasian, they have pride in construction work in what they do. Hence, the Singaporeans don't have the mentality to do the work. So the Singaporeans, they feel shy to tell people they work as sweeper, or construction work etc. I would say that our construction cost and labour cost in Singapore is one of the top in Asia. You add up all these unnecessary labour and the regulation, then we have to send our workers to training which also cost money. We have to pay these so-called "training centres" but their standards not there and they collect money like nothing. Then when they come to Singapore, still need to go for further training and thus incur more cost. The government say must need the foreign workers to go for the various training programme, all these trainings not cheap! Costs thousands in total to attend. Then at the end of the day after all these trainings, the workers tell you that they don't want to do your work. Then they "jump ship" to another company and get hired straight away.

Q15: So at the end of the day, all these circumstances that you share with me, you will consider not subcontracting more works under such tight labour conditions, meaning you won't subcontract as much as before?

A15: It is very subjective. As I mentioned earlier, if you have so many job on-hand but not enough labour, then you have to sub-out. You can secure a job but don't have enough worker, you need to sub it out, then have to pay the subcontractor. Thus you incur higher cost and it reduce your profits. No choice. Then if you don't do finish the works, the main contractor will LD (Liquidated Damages) you. It will be worse for you at the end of the day. So it really depends on your capacity, whether people trust you or whether people think your company can perform better, more reliable, financially sound, good working relationship … A lot of reasons.

6. **Would sub-sub-contracting out your firm's works to others affect your project construction productivity? How do you ensure your firm stays productive?**

Q16: Would sub-contracting your firm's work out affect your construction productivity? Any measures your firm has adopted to continuously stay productive?

A16: I think it is the same but we have to pay more to the subcontractors. If you have your own workers and you can control well, of course it is the company policy to train the workers, subject to your foreman and supervisor and how they supervise the workers. It is very subjective if you say whether productivity or no productivity. If you pay your workers well, they will work very hard right? It is the money issue. It is how you encourage your workers to do the work.

Q17: Then if you sub-contract out, then it will be another team of workers that are doing the job. Will it still affect the productivity?

A17: But you still have the supervisor to control these subcontractors. Of course they will always work faster as they are the subcontractor and would want to work hard to earn more money. It means how your supervisor control them. Subcontractor you will always talk to them but sometimes they do very fast but the quality not there. So you will be very careful to check their quality.

Q18: So that means it depends that when you sub-out your work, it is your supervisor that ensures that the work done actually meets expectations?

A18: It is very subjective. You have a good and a bad subcontractor. The good subcontractor the boss is itself part of the team and the boss is always there to ensure works are done, hence can expect better quality of work. Then some contractor firm too big, then the boss always run around then leave the job to the subcontract supervisor to do the work, then the sub-sub-contractor. Eventually end up with the quality of work not there. Then it also depends on our supervisor as well, how they control the sub-sub-cons.

Q19: So that means when you sub-out your works, you will have a supervisor that will be checking on them and to liaise with the sub-sub-contract foremen or supervisor on such work issues such as quality?

A19: Yes, it's a layer to layer approach or a level to level approach. So how to be productive?

Q20: How do you ensure that your firm stays productive? Is it via means such as financial incentives?

A20: Very hard not to say productivity. This is because there are a lot of regulation that you cannot runaway. MOM would set the rules. When you do the bearing work, you require a "fire watchmen" there. Then you need another "fireman" to do the work. Hence, do a lot of things, a lot of applications and a lot of waiting time. This is all because of too many regulations that hinders productivity. We have to follow but to follow, it's not productive anymore. In the older days, we don't care about all these regulations as there was none. But of course there is a higher risk that caused accidents. But we don't need to use so many workers. Now we have to use so many workers but doing the same work. Hence, this caused our construction cost to be very high. For Per square foot in residential, it is almost $300 per square foot in Singapore, which you can get in Malaysia for 300 Ringgit. Then in China, cheaper. Why? It is because our labour cost is very high. Do you know how much we have to pay for the levy per worker? You have not even pay your workers salary then the government take $1000 away from you. Then your workers' living quarters, one month about $300–$400. By adding up, a general worker almost cost $2000 at least. If plus overtime, an individual would cost $2000 ++ per month of cost to fork out.

Q21: In these $2000, it only consists of the levy, accommodation, and plus their basic salary but exclude overtime?

A21: Yes. So their basic pay is like a $1000++, like $30 over dollars per day their salary. Then you pay all these $2000++ and if work overtime, almost can reach $3000. Do you know how much the graduates in China, do you know how much they earned? They are getting 3000 RMB, which is about SGD$600. So you see, how much do we have to pay for our general workers? Our general workers cost more than the Chinese graduates at this point in time, hence it's a very tough time for us.

Q22: So what about the skilled workers, it definitely cost more, am I right?

A22: Yes. The skilled workers normally refer to the Chinese workers. We have to pay about SGD$3000 for them. So this include their base pay etc. It's almost our local graduate pay already. Something wrong man!

Q23: What about the Malaysian workers? In terms of skill level, they are among the best right?

A23: Still have, but very hard to employ them. They are more of the QS, higher level office staff. General workers lesser, but our production team still got Malaysians. Sometimes also not cheap. In terms of skill level, they are almost equivalent with the China workers.

Q24: Therefore, I can conclude that it is very hard for your firm to stay productive due to all these regulations that are restricting your firm's potential?

A24: We are not a standard product company. If we are such a company, we can improve on our machines, can get robots to come in and do the standard work, then this can cut down our labour. We are the design and build, every time we do a different thing. If your building like that of buildings in Europe, all standardised, then they are using the standard module to build, hence very productive.

Q25: So is it because in Singapore, there is no standardisation in our works?

A25: Yes, because every owner wants to have a different building. Every Architect wants their building to be outstanding and different. So this is not standardised. Like HDB, all use the precast, PPVC, everything also use the same thing, factory also produce the same thing.

7. **Do you have a core team of sub-subcontractors whom you always work with? Why?**

Q26: Do you have a core team of sub-subcontractors whom you always sub-out to?

A26: Yes. We do keep some of our sub-sub-contractors, just like how the main contractors keep some of their subcontractors. So we need these sub-cons.

Q27: Why is it that you need these sub-cons?

A27: Well, some are more skilled, price wise is ok and easy to control.

Q28: What do you mean by easy to control?

A28: They listen and take instructions. Because some of the sub-sub-cons their quality no good, lots of defects. So I don't want this type of subcontractors.

Q29: Are you referring to their working styles where you get to know they usually work, am I right?

A29: Yes.

8. **How do you help your sub-subcontractors to enhance their construction productivity, if any?**

Q30: How do you help your sub-subcontractors to level up in order to improve their productivity, if any?

A30: I suppose that the sub-sub-cons they should know their kind of work and know what to do.

Q31: Is it because it is their field of expertise?

A31: Yes. They should know what are the faster ways to enable productivity. The faster they work, they will make more money right? If they slow, they make lesser earnings, am I right? I should say the sub-cons that we engaged are experienced. If they no experienced, we also dare not engage them.

Q32: So you will not assist your sub-cons in improving their construction productivity as you feel that they know what they are doing?

A32: We will not interfere in how they do the work but we will brief them on the method of statement on how the work should be carried out. We will brief them on where to do first, how to do etc.

Q33: Well, upon hearing most of your responses so far, it's as though that the issues that is hindering the productivity of your firm is more linked to regulations such as those of safety? Am I right?

A33: Safety is one of the issue and the knowledge of sub-con is another issue as well. I would say that the sub-cons are not properly trained. A lot of the foreign workers they come in then start to have proper training and exposure to the works, then after a few years, then start to have the skill. Not like our ITE students, already train in their education system to have the required skills already. All these ITE students, after they study finish then also don't know where they go.

Q34: So this is because most of the foreign workers come from farming and then they come to Singapore to do construction work just to earn more money?

A34: Some for example, come to Singapore don't know how to do welding, then you have to train them as they learn from you. Then after a few years, they can do a better welding. Then you have to further train them, send them to courses and then end up we become a training centre. In the end, they "jump-ship" to other firms for better pay.

Q35: I presume this is very common in our industry?

A35: Yes, it happens everywhere. Everybody in this industry is facing the same problem, not just me.

9. **How does your firm value-add to your main contractors whom you work with?**

Q36: As a subcontractor, how does your firm try to value-add to your main contractors, whom you frequently work with?

A36: What the main-con want is responsible, qualified and knowledgeable sub-con to help them to finish their work. We know how to solve their problem, we have a team of people and we can help them to finish the work, then that's what the main-cons want right? Then also talk about the price.

Q37: So one of the issues in the construction industry is price, so what does the main-con usually do when selecting a suitable tenderer?

A37: You see, our construction industry is different from other countries' industry. Singapore industry always tender for the lowest price to give to the main contractor. One the main-con that has the lowest price is awarded the tender, they start to lose money. Then they start to cut the pay for the sub-cons. For other countries, the government they will have experts and consultants where they will draw a line like for example; $210 million which is their benchmark for a project. Tenders whose price is close to $210 million will get the job. Even higher, they will give to the closer one but they will never give the lowest one. The governments of other countries are also scared that the contractor that has the lowest tender price cannot deliver the job. That is the better system in other countries. Singapore always award to the lowest complying bid. Then the lowest main-con take the job and then they will "kill-off" the sub-cons as they cannot afford to pay their sub-contractors. So at the end of the day, the main-con go bust, cannot pay its sub-cons and then the sub-cons will "die-off" together as well. I heard now they start to improve but they still take the lowest price. But among these few lowest bids, if one of them is deemed as financially weak, because they know got a lot of court case that shows main-cons unable to pay up, then end up the firm is being sued etc. But mainly they still look at the lowest price. We do a lot of jobs for the foreign main-cons, like Samsung, all the Japanese companies as they are financially strong. Unlike the local Chinese firms, they always want to "chop us" down.

10. **If your firm is a subcontractor, what and how would you want your main contractors to help you level up your construction productivity?**

Q38: As a subcontractor, how would you want your main-contractor to help improve your construction productivity?

A38: We hope that the main-con can give us more labour and equipment such as gondola, automated lifting equipment etc. where it can cut down the amount of labour required. Normally the main-con don't do this as the main-con provides shared equip like scaffolding. Very hard as the main-con don't usually help. You want to use the equipment or machine, we as sub-cons have to buy ourselves or loan from the equipment company. This is because they view that it is the area of expertise of the sub-cons.

Q39: You also mentioned that the D&B where every project is unique where there is a limitation in employing machinery such as robots etc. Can you explain more on this with regards to productivity?

A39: For example, if all 10,000 pieces all standards. Use machine, no problem. Tomorrow I ask you do each piece differently, quite difficult right? Not standard size, hence also rather difficult to use such machinery to improve productivity.

Additional Points

Q40: What are your views regarding construction productivity in Singapore?

A40: Of course for production we do use machinery to try and support us and can help us. However, the land cost does affect the factory size, where the factory is so expensive. The cost of my factory here, if I go over to Malaysia, I can buy a factory that is the size of a production plant. My factory here cost SGD$5 million multiplied by 3. With such a price, how huge my factory can be in Malaysia… So we have limited space. Even if I buy a lot of machinery, also useless. We buy so many machine, what if no job to do, then we have so many machinery and such a big factory, then how? Then our labour is so expensive. Hence, it is very hard. We have to shift to Malaysia, Iskandar area there to set up a factory where it is of lower cost. Over in Singapore, labour and factory cost is so expensive. We want to put a lot of machine, but is it possible?

Q41: So meaning that it is the space and cost issue. Hopefully, the government can provide you with some grants to defray the cost?

A41: In terms of grants, the government have the PIC grants. Then if we can prove that after the purchase of this equipment can improve our productivity, then will be able to receive this grant.

Q42: How would you want the government then to help improve your productivity?

A42: Well, it's hard to say. You see all these safety issues, hopefully the unnecessary manpower can be cut down from 3 people to 2 people, then it will have cost savings as only need to pay 2-person salary instead of 3. These 2 guys are already safety trained, hence no need the safety supervisor. Everybody has to be responsible for their own safety, best case is to cut down to 1 person for the job. So this individual must have the safety awareness. Like this then can save more money.

All the unnecessary regulation can take away just take away as sometimes, it is really overkill.

Q43: So at the end of the day, cut down the redundant stuff, so as to make it more cost-effective and reasonable.

A43: Yes, that's right.

End of Verbatim Report 15

Annex 20: Interview Transcript 16 (Subcontractor I16)

Date of interview: Friday 11th Mar 2016.
Position of Interviewee: Managing Director.

1. **Does your firm sub-sub-contract works out to others?**

Q1: Does your firm sub-sub-contract works out to other subcontractors?
A1: Yes, we do.
Q2: What type of works does your firm do?
A2: Currently, we do a lot of sewer works, like minor sewer connections, trunk connections. For structural ends, we do those plumbing works, which is what this company started out as initially.
Q3: So your niche trade would be plumbing?
A3: yes, we started off with plumbing, then we moved off a bit more towards civil and general building. Actually, civil and general building goes hand in hand as every building has external works and that's civil works.

2. **What percentage of work does your firm typically sub-sub-contract out to others?**

Q4: May I know what is the percentage of work that your firm typically sub-contract out to your subcontractors?
A4: We usually sub-contract out around 50% only. The rest would be direct force.

3. **What types of work/trades do you typically sub-sub-contract out to others?**

Q5: What sort of trades do you sub-sub-contract out?
A5: The type of works we sub-contract out are specialist works such as water tanks for plumbing works, piping works.
Q6: So I can say that all the plumbing works, your firm will do it in-house?
A6: The main works we do it in-house would be sanitary works. Only the plumbing part we sub-out such as installation of pipes and water tanks.

L. Sui Pheng and S. Jia Zheng, *Construction Productivity in the Multilayer Subcontracting System*, Management in the Built Environment,
https://doi.org/10.1007/978-981-13-1831-3

4. **What guiding principles do you adopt to decide on (a) what, and (b) how much work to sub-sub-contract out to others?**

Q7: So for your case, any guiding principles to determine what to sub-sub-contract and how much work to sub-sub-contract?

A7: Whatever we don't have the in-house capability like the installation of water tanks, we sub-contract out. This is our number one guiding principle. The rest is when we really can't cope with it with our own manpower, only then will we sub-contract out. Therefore, we always deploy our own direct force first, see how we can cope. Only when we cannot cope, then we sub-contract out.

Q8: So it depends on your work capacity?

A8: Yes it is.

5. **In the midst of a tightening labour market and increases in foreign worker levies which has resulted in a squeeze in foreign worker numbers, do you sub-sub-contract more works out to others than ever before?**

Q9: Now under the midst of the tightening labour market like the reduction of MYEs etc., does your firm sub-sub-contract even more works out than before?

A9: No, we actually sub-contract lesser. This is because even my sub-contract can't get workers. Whenever we can, we hire directly and train them internally.

Q10: You just mentioned that you are going to hire directly. However, you also mentioned that it is hard for you to get workers right now. Does this mean that your work capacity has been decreased as well?

A10: Yes, it has decreased.

Q11: And this is a common scene throughout the construction industry I supposed and that the industry is doing pretty badly?

A11: Well, you are right that it is a common but the industry is not doing bad but slowly down.

Q12: Can I attribute to the government's tightening of the labour market that further resulted in the slowdown of the market?

A12: Um… No, it is not so much of the tightening of the labour market but more of the implementation of the property cooling measures. This is the one that contributed to the slowing down of the market and the larger economy is not doing well as well. So it is double whammy kind of effect.

Q13: And it is also made worse due to the levy as well?

A13: If you really look into it, most of the contractors are not so affected by the levy but because at the end of the day, we transfer the cost back to the client. It is that realistic as the contractors won't absorb it. Hence, construction cost is on the rise even before this slowdown.

6. **Would sub-sub-contracting out your firm's works to others affect your project construction productivity? How do you ensure your firm stays productive?**

Q14: Would sub-contracting your firm's work out affect your construction productivity? Any measures your firm has adopted to continuously stay productive?

A14: Oh yes, definitely. Because when we have to deal with our own sub-subcontractors, most of them are bosses themselves and that they are very hands-on. Hence, when it comes to documentation, they are very weak. Hence, we advise and actually have to support them in terms of permit to work and daily meetings. All these documents, they cannot handle it most of the time. Even then, it is just the paperwork, not the physical part. The actual physical work, we still need to have one or two guys to be there, based on the specifications and risk assessments and safety procedures. So we have to monitor the workers when they are doing their works based on the safe-work procedures that have been agreed prior to start of work. Hence, there are lots of eyes on them.

Q15: So there's lots of supervision required. Hence, does that increase efficiency or decrease efficiency?

A15: If you ask me, I would say it is a love-hate kind of thing. Without the sub-subcontractors, I can't produce anything, therefore no productivity as they are the specialist contractors. But they are hands-on people, they only know how to work. Hence, in terms of productivity, it will suffer a bit as my supervisors need to "eye-ball" them. So they have to leave my workers on their own. The productivity increases somewhere and then it comes down as there is a maximum bandwidth that a supervisor can do. Even supervisors are rare now as the more experienced workers has been promoted through the ranks or they have joined other bigger firms.

Q16: So overall in this sense, does it mean that your project construction productivity has decreased because of all these safety measures?

A16: Ok, like what I said that the productivity has not improved that much but when you say that it has come down, it is not true as without the sub-subcontractors, nothing will be produced. So it is a chicken and egg kind of thing. One is dependent on the other.

Q17: Hence, when you sub-contract out to your sub-subcontractors, whom are usually specialist contractors, the job will be done much faster and efficient?

A17: it's a give and take approach. Yes, it does but it sacrifices my supervisor's "bandwidth" as it is much faster in getting the job done as the specialist contractors are efficient in getting the job done at the expense of my supervisor's time, hence sacrifices something else like monitoring of these specialist contractors by my supervisors.

7. **Do you have a core team of sub-subcontractors whom you always work with? Why?**

Q18: Do you have a core team of sub-subcontractors whom you always sub-out to?

A18: Yes, we do have. In fact, most contractors have. Well it's because the working relationship has been built over so many years where some things there is no need to be said and it is expected of them to be done. If it is a new guy, then my supervisor will have to closely supervise them even more. Because most of the sub-subcontractor, they can't do documentation but that doesn't mean they are not good at their work. It just means that the way that they do their work is not theoretically safe, you get what I mean?

Q19: So that means they are not so good at their paperwork, but all along that they are doing they have been doing their trade hands-on. However, it is due to the government regulations that requires such paperwork and because they are naturally bad at paperwork, it can seem as though they are not adhering to any safety regulations and thus increases the risk of getting fine by the authorities.

A19: Yes that's correct but then again, the construction site has become increasingly complex because most of the time there are so many services running around at the same time. So not necessary because it is paper safety but because it could be a real safety issues such as I am carrying out welding to one of my water tanks and then there is some painting in a room nearby. So to the welder there is a permit that has been received and there is a go-ahead to carry out the welding. The risk is still very real and most of the time, due to tight scheduling, there can be such dangerous circumstances. Hence, it is very time-pressed in the construction site. Hence, there is also a need for someone to physically assess the site to look out for any such dangers.

Q20: Has it ever come to a point of time where your core team of subcontractors have filled to their maximum capacity and therefore, you have to source elsewhere?

A20: Mostly our sub-subcontract is mostly by introduction and referrals, we don't go by advertising. Even then, we also have to spend some time with them as we have not worked with them before and they are not sure with your working styles and hence have to sit down and let them know our expectations.

Q21: How do you actually select your core team of sub-subcontractors?

A21: The first thing would be how long they have been working with us. The next thing would be even if we never work with them before, we will check on their credentials like by making sure all their workers are certified and that the company does not have any major safety incidents before etc.

Q22: What do you mean by "how long" that the subcontracts have worked with you before?

A22: It means that the project length is not really the issue but more on how many projects that have been done together and successfully handed over before. It's more of the success rate instead of the duration.

Q23: So it is the frequency of them working with you and overtime you see whether they are reliable or not. And if they are reliable, you list them as your own core sub-subcontractors.

A23: Correct.

Q24: What are the criteria that you actually look at for your core team of subcontractors?

A24: Because we are ISO registered. We always have to update our approved list of vendors and this are requirements under ISO. So typical flags that we have get from project sites are if the contractors have not delivered their jobs on time, safety violations and workers not certified. Hence, we have these checklists….also financials etc.

Q25: With regards to financials, do you have some sort of financial standing or limit?

A25: Because most of them are really small players, hence we ask them to present their track record for us. Because their paid-up capital is really nothing and if something were to happen, they usually wind-up. Their company usually not worth much. Hence, it is more on what are the jobs that they have completed and what's the project size and we actually check. Most of them don't even do accounting as they are too small already and they only do that particular trade which is their niche.

8. **How do you help your sub-subcontractors to enhance their construction productivity, if any?**

Q26: How do you help your sub-subcontractors to level up in order to improve their productivity, if any?

A26: For us internally, we have a lot of training programmes that are mandatory. This is because of the government, we have to make sure all of our skilled workers are upgraded to be registered as core tradesmen. And if they are not able to achieve the status within this stipulated timeframe, my firm's grading actually gets affected with BCA. So internally, we actually monitor which worker is due for what upgrades and we send them for such courses.

Q27: So you actually direct your sub-subcons to actually go for such courses?

A27: Yes. Because if they can't maintain their workforce due to some lapses in training, then we are both in trouble as he lost his workforce and I lose my sub-con.

Q28: Do you actually sponsor your specialist contractors to go for training or you just remind them?

A28: We actually just remind them and ask them to go for such courses by themselves. For our own workers, we actually bond them where we pay for their training and in return they extend their service with us.

Q29: But then with regards to the levy issues and MYEs, so will that affect your manpower?

A29: With regards to foreign workers, there are a few things you need to know. Firstly, for workers that have worked for 4 or 5 years and that they have achieved the certification and skills required such as the minimum SECK (Skills Evaluation Certification Knowledge), this is a certificate that shows you have the required knowledge. If you have achieved that and they have worked in Singapore for 4–5 years, you just pay a much higher levy, double the levy in fact, but without the MYEs and this is how you retain the skillset within the company under the worst case scenario where this worker of yours is under the high levy scheme. Not all workers qualify for it. Lower skills workers they have to go back home as it is not feasible to do so as it is too expensive due to double the levy.

Q30: So that means the skilled worker has worked in Singapore for a minimum of 5 years?

A30: Yes.

Q31: And the levy for a general worker roughly cost how much?

A31: Roughly $550 per month. For those higher skilled worker, about $360 per month. Thus higher the skill of the worker, lower the levy.

Q32: What about the MYEs?

A32: MYEs are usually based on the contract sum. The main contractor is awarded the contract of $x dollars and thus is given X amount of MYEs. If the MC is fair, they will distribute it via contract figure but most of the time it doesn't happen. For example, if yours a $10 million project and got 100 MYEs, and that my subcontract value is $1 million and that I should have 10 MYEs. The main contractors always think that as plumbers, they think we don't need so much manpower, they always give more to the structural works and they don't go by contract value. Thus, it is unfair to go by the way MOM calculates MYEs. Because now is that the bigger your contract sum, especially those that goes beyond $20 million, the MYE entitlement actually shrinks as you are required to be more productive. Probably someone in MOM thinks that the worker in a multi-million dollar contract has lots of exposure in a big project and thus there is a learning curve. Theoretically, it sounds good but in reality, the same worker does not have the luxury of learning curve as in construction, scheduling of works is actually very packed. So it is not one worker having a learning curve, it is actually more than one worker each having their own learning curve at different points in time of the project. Thus, you should also look at the way how MOM issues the MYEs where it is also affecting us.

Q33: Could it be also that they have the notion that the value of the contract sum is so huge, probably you can invest in technology in aid in productivity?

A33: Probably, well all these is very academic. However, in reality it doesn't work out that way even though we hope for this as well!

Q34: Any other measures you have to aid in your sub-cons to enhance their productivity?

A34: Our in-house training unfortunately such as the development of such training is not there yet. However, we ledged on whatever training that is provided or mandated by the government. When we got BizSafe and we got the Star level, we make sure that our core team of sub-cons also achieved our same level. Whenever we can, we drag them along as well such as encouraging them to go for such courses and tell them about the benefits where eve if you don't do my job, you can still easily get others' jobs as well because of these certifications.

Q35: What about measures such as giving incentives to your sub-subcons, do you practice that?

A35: Yes, we do. We actually give rewards for achieving certain CONQUAS score, safety scores by giving them a small token sum, like a small form of monetary reward to motivate them and further encourage them.

9. **How does your firm value-add to your main contractors whom you work with?**

Q36: As a subcontractor, how does your firm try to value-add to your main contractors, whom you frequently work with?

A36: Most of the time since we are so specialise in plumbing works already, we handle the consultants ourselves and we don't deal with the main contractor's representative. So I actually free up the main contractor's M&E coordinator while we deal directly with the consultants to solve certain technical issues. Meaning that we free up your supervisor's bandwidth because we know what we are doing and he can do his own work and we will settle for you with the consultants.

Q37: That's for your specialist trade plumbing right?

A37: Yes. That's how we value-add. We be fair to the main contractors. If we can do whatever we can, we will do it to minimise inconvenience for the main contractor. Hence, it is more about work efficiency in getting the job done much faster.

10. **If your firm is a subcontractor, what and how would you want your main contractors to help you level up your construction productivity?**

Q38: As a subcontractor, how would you want your main-contractor to help improve your construction productivity?

A38: Mostly on site, the issues are mostly coordination problems. If their M&E team is not strong in coordination, then we end up fighting for a lot of coordination stuff on site itself. If the main contractor can smoothen out the coordination process on site such as using Revit software to detect clashes, even before the actual work is executed by giving us the work sequence. It will be good and that is ideal as they can tell contractors which trade will come first and that we contractors need not fight for the same workspace.

Q39: So meaning that the scheduling done by the main contractor has to be properly outlined and executed by the main contractor's project coordinated?

A39: Yes, because for the M&E works especially for most buildings, there is a main M&E trunk where you can see all the sprinklers, waste pipes all run in line together. Hence, which parties get to go first to install? If not, everybody will just rush in and do first. Hence, usually sprinkler pipes will have to done last where it will be an obstruction for the installation of the electrical wirings and plumbing pipes, hence much more inconvenience for us. Hence, maybe there is a need for main contractors to use BIM more.

Q40: So for those well-established main contractors in Singapore, they know how to use BIM as well?

A40: Yes and they expect us to know BIM as well. So far most of the projects we do all use BIM already. Maybe because we have been doing large scale projects and thus we always deal with BIM. Before the mandatory use of BIM last year, BIM is only adopted for projects of certain size, usually for large scale projects.

Q41: Is BIM well adopted in Singapore?

A41: At least for the people whom we worked with, they have already adopted BIM quite well already. Even then, the coordination is still an issue such as the scheduling of the works as the BIM team does not communicate with the project team. The BIM team just busy meeting the requirements required by the authority's submissions. I can't really blame the project team as they have sometimes very tight deadlines to meet so they just make a judgement call as BIM is an afterthought. After all, there is still shop drawings. But then, I feel it is wrong as I feel that if we can just spend more time on the planning stage regarding coordination, then there will be less clashes later. Unfortunately in reality, it is not like this. Basically, the aim of BIM is to facilitate coordination. If you can imagine, BIM is supposed to be good as if the consultants or architects have already coordinated their model, simply give to the contractors for execution. Thus construction can be so much smoother. However in reality, the architecture and structural works are coordinated but not the M&E works. Hence, M&E works is never coordinated and such coordination issues will arise due to clashes with architectural and structural works.

Additional Points

Q42: What are your perspective with regards to construction productivity in Singapore?

A42: Well, if I can give my workers information needed for them to execute their work in a manner that they can understand easily, productivity will naturally go up. Because most of the time, M&E drawings only has the schematics even before the shop drawings are out and when work has already started. It will definitely be more productive when there is some form of shop drawings rather than figuring out from the schematics as to the positioning of M&E equipment etc. Again, it boils down to coordination at the front end, then construction work will be more productive. If we can spend more time at the documentation phase of the project like creating drawings and creating coordination, it will definitely cut down a lot of cost and wastages during the construction process itself.

Q43: What about mechanisation?

A43: Oh, BCA is still trying to push for this but not very successful. They try to come out with tiling and screeding robot etc.

Q44: In your opinion, is it helping productivity or is it still not addressing the problem?

A44: Still not addressing the problem. The problem is that it is never coordinated before we execute the job. If it's coordinated, everybody knows what they are supposed to do, whether robot or no robot, everybody knows what to do based on the sequencing. It will then be very clear. If you ask me as far as worker's productivity on site is concerned, it is amazing how those wet trade guys etc., those guys are very productive already. It is amazing what they can do. Overnight you tell them that you want this done, the next morning it is done already. But why do they need to do such rush work or overnight job? If everything is coordinated at the front, the overnight work can easily be avoided. Ultimately, it is the planning that is the main issue.

Q45: What about your opinion about Singapore's construction industry with regards to the tender woes where companies undercut the tender bids of each other to become the lowest?

A45: It is not unique to Singapore as it is a market issue. Such companies take a huge risk by going very low in their tender bids and go for many other projects as well in order to sustain their cash flow and company operations.

Q46: Has there been an instance where the main contractors withhold the payments and never give back your retention sums?

A46: Yes it has occurred, even till now until today's day and age. They are very clever one as they have legal advice, especially those big firms. They know that if you take action, they know that one-third to half of your retention sums will be used for legal fees already. They will make you think twice and to them, it's of no cost as they have to hire a legal counsel anyway because of their size. So they have the resources and finance to do so. Hence, for us, we small firms suffer and get bullied by the bigger players like them. That's the game of construction. They exploit the legal framework within the SOP Act. It's good but has been exploited by the bigger players.

End of Verbatim Report 16

Annex 21: Interview Transcript 17 (Subcontractor I17)

Date of interview: Saturday 30th Jan 2016.
Position of Interviewee: Managing Director.

1. **Does your firm sub-sub-contract works out to others?**

Q1: Does your firm sub-sub-contract works out to other subcontractors?
A1: Yes, we do.

2. **What percentage of work does your firm typically sub-sub-contract out to others?**

Q2: May I know what is the percentage of work that your firm typically sub-contract out to your subcontractors?
A2: Around 50%.

3. **What types of work/trades do you typically sub-sub-contract out to others?**

Q3: What sort of trades do you sub-sub-contract out?
A3: We usually sub-out our plumbing jobs. We supply the materials to our subcontractors and they do the work, installation.
Q4: So that means you supply your product to your subcontractors to the site and they install your products?
A4: Yes.

4. **What guiding principles do you adopt to decide on (a) what, and (b) how much work to sub-sub-contract out to others?**

Q5: So for your case, any guiding principles to determine what to sub-sub-contract and how much work to sub-sub-contract?
A5: Usually, I go for efficiency in getting the job done. When you have your own workers, they will take their own time and do the works slower. However, for the sub-cons, once they finish their job, they get paid. Hence, more incentive for the sub-cons to work faster. For example, my worker can install 5 toilet bowls in a day whereas the sub-con can install 20 toilet bowls in a day. Hence, that's why I

L. Sui Pheng and S. Jia Zheng, *Construction Productivity in the Multilayer Subcontracting System*, Management in the Built Environment,
https://doi.org/10.1007/978-981-13-1831-3

sub-out. Especially for some projects that have very short time span to conduct installation works, then I sub-out. Hence, I would say that it is a win-win situation, where I can save time and money while the sub-con can earn more money.

Q6: What do you mean by "save money" when you sub-out?

A6: So an example would be that when my workers cost is $100, then they can only install 4 toilet bowls in a day. If I give sub-con $100, they can install 15 toilet bowls in a day. Hence, higher efficiency and I save money while the sub-con earns more money. Sub-con eat lunch 5 min while my workers will take an hour to eat their lunch.

5. **In the midst of a tightening labour market and increases in foreign worker levies which has resulted in a squeeze in foreign worker numbers, do you sub-sub-contract more works out to others than ever before?**

Q7: Now under the midst of the tightening labour market like the reduction of MYEs etc., does your firm sub-sub-contract even more works out than before?

A7: I will subcontract out even more. We face manpower woes as it is very difficult to get workers, all because of this quota issue. Locals don't want to do the job and when you lack the quota, you can't do much work. When we sub-out, the burden and problem rests with the sub-con, we don't need to think so much about it. So when I sub-out, it becomes a problem for the sub-cons to handle, not me. It will also help to improve productivity. Hence, higher the levy, the higher the percentage of works that we will sub-out. It's better for all parties.

6. **Would sub-sub-contracting out your firm's works to others affect your project construction productivity? How do you ensure your firm stays productive?**

Q8: Would sub-contracting your firm's work out affect your construction productivity? Any measures your firm has adopted to continuously stay productive?

A8: Sub-contracting out my firm's works will definitely affect my firm's productivity. It is not a bad thing, but on the contrary, it's good to do so as it will be even better and faster in getting the job done. This is because the sub-con will also have to be responsible for the quality of their work. If it's our own workers that make mistakes on the job, we just scold at most. However, if it is the sub-con, then they must be responsible for the work that they have done as it is all sub-out to them. Their pay can be deducted if the work they have done is not up to standard. Hence, they are more serious in getting the job done. Therefore, if the government increase the levy, then it can also aid in improving productivity.

7. **Do you have a core team of sub-subcontractors whom you always work with? Why?**

Q9: Do you have a core team of sub-subcontractors whom you always sub-out to?

A9: Yes, one or 2 core teams that I usually sub-out to. It's easier to coordinate and they know our working styles and payment mode. They know our desired quality standard and they have the credibility to get the job done. If I go for other unknown sub-contractors, I cannot sleep well at night. That's why need to have good sub-cons as the liquidated damages can be a few thousand dollars per day. Also when main contractor incurs any losses due to this LD, then we also have to pay back the main contractor. Hence, we always go for core team of subcontractors. It is only when our core team of subcontractors are too busy to handle the work load, then we have no choice but to go for other subcontractors. Sometimes, we don't even accept new projects as it is way too risky to subcontract out to sub-cons that we have never worker before. If the sub-cons not enough money, never pay their workers and their workers never come to work, also terrible headache for us.

8. **How do you help your sub-subcontractors to enhance their construction productivity, if any?**

Q10: How do you help your sub-subcontractors to level up in order to improve their productivity, if any?

A10: Well, just give the sub-cons more money, in the form of labour charge, such as higher sum per unit of toilet bowl and to tell them that if they can finish up the job within a stipulated timeframe, then you will give them an extra few percentage of payment. If it is not done properly, then you deduct certain percentage off their pay. So it is a system of reward and penalty. So for example if the sub-cons can complete the job within 1 week, I pay them an extra 20%. Then if can complete within 2 weeks, pay them an extra 10%. If cannot complete the job within the stipulated time frame of 1 month, then I deduct 3% of their payment. This will definitely help spur productivity.

9. **How does your firm value-add to your main contractors whom you work with?**

Q11: As a subcontractor, how does your firm try to value-add to your main contractors, whom you frequently work with?

A11: We strive to deliver our job on time so that we do not cause our main contractors to get LD. Once they get LD from the client, it is also detrimental to us as the LD is also imposed onto us. Our payment gets deducted as well. So it is important for us to deliver our job on time with good workmanship and quality.

10. **If your firm is a subcontractor, what and how would you want your main contractors to help you level up your construction productivity?**

Q12: As a subcontractor, how would you want your main-contractor to help improve your construction productivity?

A12: Probably more on coordination. Because without proper coordination, very sure can lead to cock-ups, unnecessary delays, miscommunication and worse still, misunderstanding. Therefore, the main contractors should sit down with their sub-cons and discuss the scheduling with them to avoid any potential pitfalls or clashes.

Additional Points

Q13: What are your views regarding construction industry in Singapore?

A13: I feel that the construction industry in Singapore will continue to decline and shrink in the next few years. Now even government projects that have been tendered out successfully, are done by main contractors that are doing these projects at a loss. Main contractors do this in order to survive through such difficult times, especially when the market is not doing well and they still have workers to feed. They need jobs to survive and sustain for the time being. These main contractors still need to pay off the high cost of the levy if they still want to retain the foreign workers due to obligation issues. Even though many firms undercut each other's tender bid to get the job, they are very willing to make a complete loss in their operations so as to enable them to have sufficient funds through these jobs for them to tide over. Sometimes when they undercut too much, it cost these contractors to go bankrupt. Just like those HDB works in Kallang, Sengkang and Punggol, the main contractor just go bankrupt only, because there is no need to go to jail anyway. This is because there is way too much supply (Contractors) as supposed to demand (Government Projects). This is why such cycle is very vicious which is very unhealthy for the construction industry. When market good, everybody earns. When market is bad, everyone just scramble for jobs.

Q14: So this has somewhat affected construction productivity in Singapore?

A14: Definitely has affected construction productivity. How can one be productive when you're trying to survive in such bad times? Also due to some "unwritten" industry practice that is common in Singapore. For example, a typical main contractor has 4 subcontractors. The main contractor will never-ever finish paying the subcontractors where the main contractor will still hold back 10–20% of the sub-cons payment. Some of the main contractors are even more ruthless, hold back up to 50%. These main contractors in a bid to gain control over these sub-cons will resort to such lowly tactics. In fact, when the main contractors have their annual dinner and dance, they make most of the sub-cons pay for their events like these. The main contractors will only pay back the next 20% of their sub-cons' payment in the next project with them and this cycle continues on and on. So in that sense, that's why some of the main contractors are willing to undercut their tender bid to get projects after projects, all in a bid to lower their costs and maintain their cash flow by keeping their sub-cons under their control. This is what I call the "construction game" in the industry. Don't believe me, just ask any sub-con, they will tell you that their main-cons never pay them back their retention sums even after the project is done. In addition, that's also what sub-cons fear—No jobs in future from the main-con, especially when the economy is bad, should they always claim on time from the main-cons or take legal action to get back their payment from the main-con. Also legal action fees are too expensive for sub-cons to afford. So that's why sub-cons will always get bullied by the main contractors, let alone accrue construction productivity gains for them.

In addition, there is a dilemma for contractors as well. If they do their work honestly, by bidding honestly, they will be the most expensive amongst all other bids, which is undesirable. That's why I say construction industry not an easy place to be. Even if sub-con want to get their payment back, have to go for karaoke and foot the bill, where majority of the expenses is by the main contractor's guys (also known as project manager etc.) It is a tough industry indeed.

End of Verbatim Report 17

Annex 22: Interview Transcript 18 (Subcontractor I18)

Date of interview: Saturday 12th Mar 2016.
Position of Interviewee: Managing Director.

1. **Does your firm sub-sub-contract works out to others?**

Q1: Does your firm sub-sub-contract works out to other subcontractors?
A1: Yes.

2. **What percentage of work does your firm typically sub-sub-contract out to others?**

Q2: May I know what is the percentage of work that your firm typically sub-contract out to your subcontractors?
A2: Around 70% of my overall trade and value.

3. **What types of work/trades do you typically sub-sub-contract out to others?**

Q3: What sort of trades do you sub-sub-contract out?
A3: We usually sub out specialist works, meaning doors, windows, ceiling, M&E etc.
Q4: So what is the nature of the work that your firm undertakes?
A4: We do mainly construction of buildings, finishes, etc. So we do the building from scratch.
Q5: So what sort of works that your firm does in-house?
A5: Work that is more complex, like in-house plaster, tiling, brick-laying, screeding. Our main task is to mainly support such works like minor works such as touch-up etc. So basically, we will touch up and ensure that the works are finished completely.
Q6: So your main task would be to support, in the sense that there is adequate coordination and the supply of the equipment?
A6: Yes. We are currently doing lots of residential projects, so when we do residential, especially if it's government projects, have to tidy up very nicely. Therefore, we need very close-up supervision with regards to all these.

L. Sui Pheng and S. Jia Zheng, *Construction Productivity in the Multilayer Subcontracting System*, Management in the Built Environment,
https://doi.org/10.1007/978-981-13-1831-3

4. **What guiding principles do you adopt to decide on (a) what, and (b) how much work to sub-sub-contract out to others?**

Q7: So for your case, any guiding principles to determine what to sub-sub-contract and how much work to sub-sub-contract?

A7: Subcontracting is basically like the specialist works where they are more specialised. You cannot train your workers to be so specialised as you don't have these kind of works every day. So come to specialist works like timber flooring, the work involve is for these specialist, they everyday do such works for every projects. They can give you better price and they know their relevant contacts, assuming you undertake their services. The reason very simple because they have been doing this trade every day and everywhere they go and every contractor they supply to and they have better control of the work. Whereas, we cannot pay so many people and thus we can only support them.

Q8: In other words, that means you sub-contract when you don't have the necessary expertise.

A8: Yes. We will still have a small strength of tilers even though we sub-contract out. We have our own people, let's say that if the contractors cannot perform, then we have to go in and tidy up to ensure that the project is not left undone.

Q9: So meaning that you still have your own in-house workers?

A9: Very small part I would say, where these workers can multi-task between different trades.

Q10: Oh, you mean multi-skilled workers, am I right?

A10: Yes. In addition, due to time concerns where sometimes the specialist contractors will not do for you as it is just simply minor touch-up. Hence, still need to have own in-house workers that can multi-task between different trades so as to save time.

Q11: So to recap, the guiding principles that you adopt for subcontracting are due to lack of expertise, but nonetheless, you still have your own in-house staff to ensure proper completion of the work.

A11: It is also due to competitions where you have to get the best specialist contractors but no adequate or specialised machinery to do the job you see… it can't be that you have the best specialist contractors but don't have the machinery or equipment for them to do their work right…

Q12: So it means that subcontracting is a much more efficient and cost effective to sub-out the works?

A12: Yes, it is.

5. **In the midst of a tightening labour market and increases in foreign worker levies which has resulted in a squeeze in foreign worker numbers, do you sub-sub-contract more works out to others than ever before?**

Q13: Now under the midst of the tightening labour market like the reduction of MYEs etc., does your firm sub-sub-contract even more works out than before?

A13: For us, I think it is the same as before. We do feel a little bit of pinch. If the industry has more projects, then surely the sub-subcons can have more work or vice versa like now since the economy is bad. One will affect the other.

6. **Would sub-sub-contracting out your firm's works to others affect your project construction productivity? How do you ensure your firm stays productive?**

Q14: Would sub-contracting your firm's work out affect your construction productivity? Any measures your firm has adopted to continuously stay productive?

A14: Certainly yes. The specialist contractors will do their specialist work as your firm does not have the necessary expertise to do the work. Hence, you sub-out to them. These specialist contractors have the necessary manpower, skills and tools to do the job and thus, they can give you more competitive advantage. However, if it is a simple job, then no need to sub-out, then your firm will be more competitive than the specialist contractors.

7. **Do you have a core team of sub-subcontractors whom you always work with? Why?**

Q15: Do you have a core team of sub-subcontractors whom you always sub-out to?

A15: Yes, they are called our regular subcontractors. They have work with you more often and they know your working style etc. However, the disadvantage is price. Sometimes, upon checking the price, their price can be higher than others and it doesn't make sense. So for example, 50% I use the same while the remaining 50% I have to search for others.

Q16: So that's how you select your desired subcontractors for the job even though you have a core team of subcontractors. So the percentage that you have just given, is it your evaluation criteria for your sub-cons?

A16: I would say more on cost. This is because the list of contractors from your core team of subcontractors are my regular sub-cons. When I say regular, that means these people are all ok and they are all competent to do for you. You may select anyone from the list of regular sub-cons but also compare their prices. If too steep, we will pick another sub-con whose price is lower.

Q17: When you select new sub-cons for your jobs, is there a criteria that you look out for?

A17: Yes, of course. You have to see whether they are competent and whether the work requires technical expertise. In addition, they have to know the product that they are selling to you and so you have to make sure that you have met the specifications.

Q18: So meaning that at the end of the day, when you search for new sub-cons you have to look at their competency level and necessary expertise.

A18: Yes, it is like this.

Q19: The background of the firm as well as its financial standing, do you factor that in when selecting your sub-cons?

A19: Yes, we do factor that in to a certain extent. This is because most of the time, a lot of things are based on trust. You can ask them if they face any problem and from there, you can tell how big their work capacity is and whether can they get the job done.

Q20: So you look for trust more than the financial standing?

A20: I would say that based on your own intuition where you question them and you can see them and you can roughly know what kinds of jobs they are currently doing. If it is a bigger job and they have done a lot, then I think competency wise is not a problem. This is because the next thing they have to meet is your specifications. Actually, they must hit the specifications.

Q21: Oh yes, what about the paid-up capital?

A21: I think for subcontracting you usually don't need that, because what we undertake, the project size is not big. Especially good where they have done this many times that forms their track record. So you need to be aware of all these, especially so if the sub-con is new.

If the new firm's finances not very good but workmanship very good, then I can also think of a way to cooperate with the new contractor, such as by supporting them financially or aiding them in procuring the necessary tools and equipment.

8. **How do you help your sub-subcontractors to enhance their construction productivity, if any?**

Q22: How do you help your sub-subcontractors to level up in order to improve their productivity, if any?

A22: I think presently it is a bit difficult. We do have our system and planning part, especially before you do must plan your work, make sure that everything is drawn properly before you carry out. In addition, there are also criteria to control the quality etc. So it is a bit difficult for our local subcontractors to address all these things. We still don't have the people who are willing to build up their company, where I am talking about very proper way, good way for long term where there are benefits. You got to plan the work, you must have adjustment, good paperwork such as asking for confirmation. Therefore I think for local subcontractors, it is very difficult to achieve all these things as it also boils down to cost such as improvisation. They simply can't afford.

Q23: So can I say that sub-cons they don't really bother too much on coordination?

A23: They only do the practical way, simple way and cheap way. But to ask them to do the proper way, means that actually you really need to sit down, study and draw it out properly and do proper documentation. So sometimes, when these sub-subcons don't have that kind of expectation, we will be the ones that do it.

Q24: Any other measures or initiatives that your firm has done to help your sub-subcons?

A24: Hmmm… I would say more on the detailing and technical aspects of the drawings. If you want to hire the more organised sub-subcontractors, then it will cost more. If not, hire the less organised one where they can get the job done but unable to do the proper and necessary procedures and paperwork.

Q25: Hence, so at the end of the day, cost is still a huge issue?

A25: Yes, it is. In fact, it is the main issue.

Q26: So what about financial incentive, would it help?

A26: In our case, it is actually very rare as the type of projects that we do are mostly on high-end landed properties like bungalows or mansions. So usually the owners or quite reserved in giving such financial incentives.

9. **How does your firm value-add to your main contractors whom you work with?**

Q27: As a subcontractor, how does your firm try to value-add to your main contractors, whom you frequently work with?

A27: We will ensure that our job is done on time so that we do not cause our main contractors to get LD. This is because if the main contractor get LD, then it will affect our earnings as our payment gets deducted as well.

10. **If your firm is a subcontractor, what and how would you want your main contractors to help you level up your construction productivity?**

Q28: As a subcontractor, how would you want your main-contractor to help improve your construction productivity?

A28: Coordination is a big issue. Because without proper coordination, it can lead unnecessary delays, miscommunication etc. Therefore, the main contractors should sit down with their sub-cons and discuss the scheduling with them to avoid any potential clashes as much as possible. But you know, so far what I've seen, such practice will not be carried out in the near time as everyone is just rushing for time in construction projects.

Additional Points

Q29: What are your views on construction productivity in Singapore?

A29: For landed projects that we are currently doing, I think it all boils down to design. So productivity depends on the design. Every architect is looking for new designs for their buildings and they will not build a house where it has the same design. They would rather go for a unique design, thereby making the project very complex. This will in turn make construction very difficult. This can compromise on safety as well as where the scaffolding must be assembled and dismantled for each different trade. Then, this makes it very tedious and time consuming and bear in mind that there is no extension of time. Therefore, productivity is very difficult to measure.

Q30: Lastly, with regards to progress payment issues, what is your opinion on the current state of the sub-contracting practices in Singapore? Is it also causing the negative impact to our local construction industry?

A30: Sub cons they always undercut each other. The main contractors always cut from sub-cons say around 5%, in order for main contractor to break even and thus, it becomes a vicious cycle.

End of Verbatim Report 18

Annex 23: Interview Transcript 19 (Subcontractor I19)

Date of interview: Saturday 9th Apr 2016.
Position of Interviewee: Director.

1. **Does your firm sub-sub-contract works out to others?**

Q1: Does your firm sub-sub-contract works out to other subcontractors?
A1: Yes, we do. We are the main sub-contractor.

2. **What percentage of work does your firm typically sub-sub-contract out to others?**

Q2: May I know what is the percentage of work that your firm typically sub-contract out to your subcontractors?
A2: We subcontract out roughly about 50% of our works.

3. **What types of work/trades do you typically sub-sub-contract out to others?**

Q3: What sort of trades do you sub-sub-contract out?
A3: Timber flooring for small individual units like HDB or in condominiums. For large projects, it is done by us.

4. **What guiding principles do you adopt to decide on (a) what, and (b) how much work to sub-sub-contract out to others?**

Q4: So for your case, any guiding principles to determine what to sub-sub-contract and how much work to sub-sub-contract?
A4: I would say the lack of time and manpower. Let's say I have a project for 10 blocks of HDB, then I also have another 10 blocks of HDB. Then happen to be in one work package. As such, we will have to increase manpower and this is the time we have to look for subcontractors. Otherwise, we also don't have much manpower to do. The reason is that it is more cost-effective to sub-out my works under such circumstances. If the job has a longer timeframe, then it is ok for us to accept the job that has a timeframe of 2–3 months. However, if the timeframe is too short, say like the job suddenly comes in and wants us to finish within 1 month, then we have time

L. Sui Pheng and S. Jia Zheng, *Construction Productivity in the Multilayer Subcontracting System*, Management in the Built Environment,
https://doi.org/10.1007/978-981-13-1831-3

shortage to deliver it. So we sub-out. Next, to compensate the time back. Otherwise, we would not sub it out.

Q5: So usually when the main contractor approaches you, then it is usually quite rush?

A5: They already might be trying to finish within their deadline despite getting delayed earlier as stipulated in their contract, say for example eight months. So they already delayed in the other three months. By right the earlier 3 months, we can meet the target. If the main contractor is late for 3 months, I got to think of ways to fulfil the 3 months back. The only way is to increase manpower. If we are given 3 months, we can finish the job. Normally, government jobs or whatever jobs have such deadline and must handover if not get liquidated damage. But the main contractor must also ensure that they rush the job despite the earlier delays. If they can't handover or handover late despite me rushing my job within the much shorter timeframe but yet cannot finish the job for the main contractor, then it is not our problem. This is because you have to work day and night to meet the shorter deadline than usual. So this is the point that we look for sub-subcontractors. Mainly because of time constraints.

5. **In the midst of a tightening labour market and increases in foreign worker levies which has resulted in a squeeze in foreign worker numbers, do you sub-sub-contract more works out to others than ever before?**

Q6: Now under the midst of the tightening labour market like the reduction of MYEs etc., does your firm sub-sub-contract even more works out than before?

A6: Yes, we do. Not subcontract more out but within our means. 50% or less. It will not be more than 50%, roughly around 30–40% of the works we will sub-it out. It depends as sometimes we take less jobs. We try not to sub it out as much as possible. Actually, all the while it has been still the same as these are skilled workers, where they undertake manual work, thus we cannot sub all out to machine. Finishing works we have to do on our own as it is a specialist trade and newcomers that come in and will have senior colleagues to train up the new staffs.

Q7: so you are heavily dependent on the manpower?

A7: Yes, it is. However, the levy problem is a bit high. We try to also go for core trade also. One is installation while the other is sanding. We have fulfil all these. Of course, when the new guy comes in at least must go for 1–2 years of training to get the necessary skills, have to go for all these training. But important is the worker wants to come back and work for us again. We will definitely apply for him to come back again because he has learnt from us for 3–4 years and have the necessary skills. What we scared is that, which my point is, let's say we engage a foreign worker in, already worked for me for 3 years. Then the worker find other place where there is better salary. This is the big headache for us as training the workers are just so hard. Then just because of other company employs them as they have 3 years' experience, they are considered very experienced already. The other side just pay a little bit more for the workers, it's worthwhile for these companies. Our side not worth. So this is the problem and other companies just offer slightly more then take our former workers whom we train until everything ok already. Also, the worker already knows where he wants to go, so back to square one, have to conduct retraining again then the worker leave again after a few years then we retrain new batch again. A cycle you see. Every contractor is also facing this kind of problem.

6. **Would sub-sub-contracting out your firm's works to others affect your project construction productivity? How do you ensure your firm stays productive?**

Q8: Would sub-contracting your firm's work out affect your construction productivity? Any measures your firm has adopted to continuously stay productive?

A8: We have supervisors to look after the workers, also known as foreman etc. to look at productivity, to ensure productivity will hit a certain percentage. This one I think it is ok as you have the supervisor to ensure this and to look after. Definitely, if you sub-contract out to sub-subcons, it will be faster one. Even faster than our own workers, 100%. This is because they want to make money. When we sub-contract out, we don't care whether how much work you have done in a day. If we can do 2 rooms, they can do 6 rooms. They are very skilled. Work wise for these subcontractors still ok, if compared to our standard, a few percentage standard difference. That's why finishing we do our own as we don't let them do it. They will do the first part while we do the ending.

Q9: Is it because finishing is your core trade?

A9: Yes, it is. We call it the specialised trade. The subcons can do, but they do already you sure have to come back to do touching up.

Q10: Because your side is particular about quality?

A10: Yes, unless you pay him at the rate of what the subcons want, then they will do it even better.

Q11: Then if they do it very fast and efficient, is the quality still acceptable?

A11: It is still acceptable, but only thing is that finishing cannot make it. They do the first 2 parts, such as sanding or sandrub etc, then it will be ok.

7. **Do you have a core team of sub-subcontractors whom you always work with? Why?**

Q12: Do you have a core team of sub-subcontractors whom you always sub-out to?

A12: Yes. Because they know our standards. If no good, they know. So they are familiar with my working style and the competency required to get the work done.

8. **How do you help your sub-subcontractors to enhance their construction productivity, if any?**

Q13: How do you help your sub-subcontractors to level up in order to improve their productivity, if any?

A13: This one you no need to help them. They automatically know how to do it faster and more efficient. Sub cons good in this part as they are always faster than my workers. They are concerned about money and that's what made them productive.

9. **How does your firm value-add to your main contractors whom you work with?**

Q14: As a subcontractor, how does your firm try to value-add to your main contractors, whom you frequently work with?

A14: This one normally the main-con will have the trade meetings where they will come to help us. A good project manager is very important where it can help in coordinating the works properly and organise properly. If you handover to us in time, with good procedure then we follow up very smoothly. Then less hiccups, less screw up, more consistent thus not affecting productivity. Main thing is that they have sound arrangement. They are ok, we are ok.

10. **If your firm is a subcontractor, what and how would you want your main contractors to help you level up your construction productivity?**

Q15: As a subcontractor, how would you want your main-contractor to help improve your construction productivity?

A15: It's all about the management. But more important is the progress payment. I think now is better. HDB is involved in this payment issue.

Q16: So meaning that work management as well as coordination from the main-con with the sub-cons is very important. With better coordination, things can be done faster and with higher productivity?

A16: Yes. Every trade must know who come first. Then if there are problem, must speak up and then resolve the issues and adjust. I think it's best to penalise errant contractors who never finish their trades within the stipulated time. Your deadline is when and where and if due to negligence by these sub-cons, can cause main-con problems. Main-con receives problems because of the sub-cons, then it's best for main-con to penalize the sub-con via warnings before initiating LD on them. That is why communicating is important.

Q17: So this is to prevent clashes of the work trades as well?

A17: Yes.

Q18: The other thing you talk about is progress payments. There are several sub cons and sub-sub-cons not getting paid by main-cons.

A18: Well, it depends on the contract usually.

Q19: Since the market isn't doing too well, have you heard of any instances where the main-cons actually withhold the progress payments to the sub-cons?

A19: There is such an occurrence. But so far we are pretty lucky not to have met one. The market is open, so the moment they do this, they will be affected by bad reputation. HDB will also step in should there be any progress payments issues with regards to any part of their projects. However, this option is not available for private sector projects. Sometimes for these private sector projects, it is not so much of the developer's fault but in fact it is the main-con's fault for holding back the progress payments. Developers usually pay on time.

Additional Points

Q20: What are your views regarding construction productivity in Singapore?

A20: I think it is quite good now. The team management is quite good where any issues they bring out timely. Now the problems are mainly due to schedule and discussion problems.

Q21: So once scheduling and sequence issues have been ironed out, then there is definitely going to be improved productivity?

A21: Yes. Because of the machinery as well and that the government has also subsidised where it encouraged contractors to adopt mechanisation. I pay 50% and government pays 50%.

Q22: For now you see in Singapore that there are improvements in construction productivity?

A22: Yes.

End of Verbatim Report 19

Annex 24: Interview Transcript 20 (Subcontractor I20)

Date of interview: Friday 15th Apr 2016.
Position of Interviewee: Managing Director.

1. **Does your firm sub-sub-contract works out to others?**

Q1: Does your firm sub-sub-contract works out to other subcontractors?

A1: Yes, we do. We mainly subcontract out specialist works such as M&E works which we don't have the skills and knowledge to do. But M&E works, having said that, for security, we do our own and we don't subcontract out. For example, the electrical security installation, we do it in-house.

Q2: So meaning like the CCTV installation, alarms etc.?

A2: Yes.

2. **What percentage of work does your firm typically sub-sub-contract out to others?**

Q3: May I know what is the percentage of work that your firm typically sub-contract out to your subcontractors?

A3: It is about 40%. However, it also depends on the nature of the contract. Some contract about 25%. For M&E works, it is only about 25%.

Q4: Then the 40% is for what sort of trade?

A4: It's for bigger projects like more M&E works and specialist works required like aluminium and glazing. That one we do not do by ourselves. But all other works like timber works we do by ourselves.

3. **What types of work/trades do you typically sub-sub-contract out to others?**

Q5: What sort of trades do you sub-sub-contract out?

A5: We usually sub out specialist trades. General contractor trades we do it by ourselves.

Q6: So generally, your firm is a general builder?

A6: Yes, general builder. And we are a small contractor. We have only about 80 people so we are a small contractor.

L. Sui Pheng and S. Jia Zheng, *Construction Productivity in the Multilayer Subcontracting System*, Management in the Built Environment,
https://doi.org/10.1007/978-981-13-1831-3

4. **What guiding principles do you adopt to decide on (a) what, and (b) how much work to sub-sub-contract out to others?**

Q7: So for your case, any guiding principles to determine what to sub-sub-contract and how much work to sub-sub-contract?

A7: I think we firstly, are doing a business. We are looking at profitability. So if we think that certain work we can sub it out and can have bigger margin, then we do it. But we are very particular on the sub-con's level of competency. Normally, we exercise something called verification of competency (VOC) before we award the job to this sub-contractor. And I tell you that in this line, usually we have a responsibility to handover a job or project nicely to our client. In terms of quality, control must be there as well as future maintenance, must be there. So we are concerned about the works done by our subcontractors, whether they are up to standard or mark. So for subletting most of the time, we try to avoid unless it is necessary. But I tell you from the business angle, usually it is faster and gives better margin. If you use your own workers, productivity is slightly lower because they are not specialist. Having said that, in this line a lot of time, when you use your sub-contractor, you still need to support them. It is where something easy to do, they will try to finish very fast. When it comes to corners or very detailed parts, they always try to avoid. When you keep calling them and they don't come back, then you have to do it yourself. So I always believe that we ourselves, must be competent in all aspects. Even though we cannot do all, come to ending, we must be able to do it.

Q8: So this is in the case where you sub-out your works to your sub-subcons and then they find all sorts of excuses not to complete?

A8: Not like that. They will go in let's say for this area, we got specialist works like special flooring. They will finish the whole lot, then claim 90% from you and then another 10% because certain works cannot be done at the same time, you have to wait until other trades finish, then you can come in and finish it for completion. Then very hard to get them to come back. Because they have already earned the 90% from you and that's the problem with the market currently. That's why all the time, we must work with people that we know, such that they know that's our way of working and they will come back you see. From time to time, you will meet specialist contractors that have a lack of responsibility. You meet some bad hats and it's unavoidable, part and parcel of this industry.

Q9: So the specialist works that you subcontract out is like flooring and tiling etc.?

A9: No, that one normally we do our own unless it is a special coating or covering. Mainly we sub-out M&E works and specialist trade like glazing, glass works.

5. **In the midst of a tightening labour market and increases in foreign worker levies which has resulted in a squeeze in foreign worker numbers, do you sub-sub-contract more works out to others than ever before?**

Q10: Now under the midst of the tightening labour market like the reduction of MYEs etc., does your firm sub-sub-contract even more works out than before?

A10: I don’t think that has any impact on us. This is because, our staff retention is very high. I think levy, there is no U-turn, according to the government. So we just maintain our same workforce. But some contractor there is some impact, especially if they have no jobs in hand. They let go of their workers or downsize. I think that we heard in the market quite a lot, because downsizing. All because of lack of contracts. I think it is sometimes, the way I looked at it, I don’t know how to explain. A lot of people say no job, but whenever you have jobs, cannot find people. True! Exactly, as they cannot find people as people are downsizing. This is the imbalance, as you don't know when it is enough. It is called the Hobson’s choice, you go also die, you don’t go also die.

6. **Would sub-sub-contracting out your firm's works to others affect your project construction productivity? How do you ensure your firm stays productive?**

Q11: Would sub-contracting your firm’s work out affect your construction productivity? Any measures your firm has adopted to continuously stay productive?

A11: I think for productivity versus good workmanship and we have more things to beat in the market now, especially like the regulatory requirements on safety. We are so-called at the infant stage, compared to the UK where their workplace safety and health already practiced for more than 40 years. In Singapore, only started for less than 10 years. So it is still in their infant stages. A lot of people would have to adopt good practices in the building, thereby causing productivity to go down. Safety is paramount, as you take care of safety first, then indirectly, will slow down productivity, all because you do not know how to practise safety correctly and in the best manner. So you have to take time to adopt good practices and your workmen also need to be trained until they are skilled enough and fast enough to react to the safety requirements through proper planning. So planning and time factor will also affect. So whether we sub-let or we do it on our own, at the end of the day, the quality control is in our hands. We don’t see that we should sacrifice or compromise on the quality just because we want productivity. We always look at construction where we do houses. We look at it as more importantly, as an art piece, we cannot just look at the commercial value and live with it. So a lot of times, we have happy customers. So in terms of quality control is that we have to start on our own, not rely on architect and consultants to spot your defects. So whatever we do, we have to do our best!

7. **Do you have a core team of sub-subcontractors whom you always work with? Why?**

Q12: Do you have a core team of sub-subcontractors whom you always sub-out to?

A12: Yes, we have. The reason is simple, they understand our standards and they understand our requirements. So less teething problems and can get along with each other better. And then when I go there when the subcon did something wrong, I just need to state what he did wrong and the subcon knows my standard and he will straight away rectify. I don’t want to spend half an hour talking to him and that’s not what I want. Hence, having a core team of subcons is very important as using the same sub-con, sometimes more costly for us. Because they know our standards, then they will abide by that standard. At the same time, they are very firm on their pricing. At the end of the day, to us, must be win-win. I personally believe this.

8. **How do you help your sub-subcontractors to enhance their construction productivity, if any?**

Q13: How do you help your sub-subcontractors to level up in order to improve their productivity, if any?

A13: Nowadays, we go through a lot of mechanisation, so we ask our contractors to buy more new equipment, more for productivity and at the same time, conduct Continuing Education and Training (CET). So we tell them that they have to up their skills and raise their bar, so a lot of things we have to find ways how to improve productivity and quality through continuous skill training and we also realise a very important thing. Last time, we don't give so much time for preparation before implementation. So I just pass you the drawings and then straight away, you go to work already. So I say no. This time we have ample time to go through, kick-off meetings, understand what I want, understand what you want, understand what materials you must procure first, even though I have not given you any award yet but I can tell you straight away to procure the materials. I will follow up with the letter of award. This is because there is a layer of trust in between these areas and thus there is no problem. By doing so, I think they will increase productivity and also cut down their costs and then we also give them money upfront to help them with the cash flow.

In addition, we have a meeting first, talk to them to ensure that they understand their scope of works and at the same time, the scope of works of others. Because in construction, it is not you do your own work while the others do by themselves. Not work in silos. Rather, everybody must work together. So in terms of sequencing, you have ample time to prepare. At the same time, we also help them to procure materials early. Because if you don't procure the materials early, then the sub-subcons will say that they cannot buy the materials required for the job. Then I also got headache. Then because no materials cannot finish the work, then result in delay.

9. **How does your firm value-add to your main contractors whom you work with?**

Q14: As a subcontractor, how does your firm try to value-add to your main contractors, whom you frequently work with?

A14: We strive to deliver our job on time so that we don't get LD (Liquidated Damages). It is detrimental to us as the LD is also imposed onto us. Our payment gets deducted as well. So it is important for us to deliver our job on time with good workmanship and quality to the satisfaction of the main-cons.

10. **If your firm is a subcontractor, what and how would you want your main contractors to help you level up your construction productivity?**

Q15: As a subcontractor, how would you want your main-contractor to help improve your construction productivity?

A15: I would hope that there is more networking sessions to value add to us. It would be better to have a networking platform such as bringing in the key players in the construction industry such as the MOM, BCA etc. to let us know what the ministry is doing, then what the newly adopted changes in regulations are and also sharing of best practices. So in future, these are some of the steps to take if one of the subcons wants to grow their firms to make sure that you work towards this direction. At the same time, it would also be great if they can introduce mapping technologies, tools and mechanisation through these networking sessions by bringing in suppliers etc. to let us know whether it is feasible to adopt these technologies that are promoted by BCA and MOM.

Q16: So in a way, what you hope for is some avenue to help firms like yours improve your operations?

A16: Not only that, it is good to provide a social platform to share notes on how to solve various construction issues, difficulties, woes like LinkedIn. Sometimes, can also try to share information like certain clients who are bad paymasters. So that when firms like us enter their "arena", we have to open our eyes.

Q17: With regards to the subcons, usually what are their worries and concerns?

A17: Their concern is cashflow, where their cashflow can be very different. When you are talking about the very small contractors like sub-cons, cashflow is really a lifeblood for them. Then secondly is on worker's capability. For these subcons, their staff retention is not there. So sometimes the workers work for a while there and when they are familiar with the work, they jump ship to better firms. I see most of the sub-cons face the same problem is that they lack of planning before implementation. They tend to take too many jobs as they are specialists and not every job is the same and thus this is a problem. So lack of preparation or understand the contract of the scope of the works. So when they don't understand, they don't plan. So this results in full of surprises when they go to the site. To improve productivity is firstly, to have a mindset change. You must make sure that he is ready to accept changes and ready to adopt new practices, ready to take advice, ready to learn and ready to raise the bar in terms of improving quality.

Q18: So are you implying that there is a lot of resistance to this mindset change?

A18: There is a need to raise the level as an entry level as a contractor. You just need 1 hour to register the company and then you can do contracting already. For those people, when the entry level is low, they come in and start to grab jobs at any cost and then they spoil market. So we need to raise this bar like security and manpower companies need to have at least $300,000 before they can register their firms. For construction no need for this, simply $50 will do and then can register the firm already.

Q19: So this will slowly kill off the industry due to all these bad sheep?

A19: Yes.

Additional Points

Q20: What are your views regarding the construction industry in Singapore?

A20: I take issues with the multiple layers of subcontracting practices that are occurring in our industry. I am not supportive of multiple layers of subcontracting. When I sub-out, I make sure that I only sub-contract to one final sub-subcontractor. If they insisted on subcontracting more works down the line (also known as other sub-sub-subcontractors) and we discover them having done so, I will never work with them again. Not only I do this because of quality control, but you see if this is allowed to continue one or 2 layers down, the guy below, the price that he gets because everybody get a cut, then he will try to cut corners as well. If not, he cannot make money. I would rather take the more painful approach where I don't care about my sub-sub-contractor, I pay you this price, I make sure you do on your own. If you cannot do on your own, I rather go direct to the sub-sub-subcontractor. Then, I give him the bottom guy a better price.

Q21: So in our industry, usually it is quite common for this multiple tiers of subcontracting to occur?

A21: No, usually it is more for bigger projects. Hence, usually for our sub-sub-cons, we select using the VOC approach which I mentioned earlier. It's something like a pre-qualification test. Because if we don't select, in the process of the project, if got personal issues like cash flow issues, if small we can still help him. If big, due to other projects, you got headache already as you cannot focus on your job. Sometimes when the sub-sub-con only has 10–15 workers then he go take up 10 jobs, more than he can stomach. Some of the contractors are very greedy, job come only they take, job done already they run. All these without planning. Some of them they feel that if they don't take the job now, once economy turns bad, no jobs come. But construction is like when we eat lunch or dinner where we have only 1 stomach. You eat the food till you full obviously you cannot take up any other jobs already right? So this is a very similar example. So I say that these people can talk only, say they can finish the jobs but I don't believe them. Sure cannot deliver the job.

Q22: So the preparation stage go through with all the parties?

A22: Yes, the preparation stage will go through with our sub-subcons and it will be done by my project team.

Q23: So this multiple tiers of subcontracting practices has somewhat affected construction productivity in Singapore?

A23: For this part, personally, I don't really comment on much. So long as they can deliver and in between definitely got people suffer, that I can tell you but don't know which layer. At the end of the day, so long as the main con is happy, client is happy and end product is acceptable, I can't really comment much on this. But in general, I don't want to practice and adopt this.

Q24: Is it due to the risk?

A24: Not only about the risk, we must engage with the people who do the work, not agents. You see that if you go by layers, you are actually working with agents. Whatever has been talked at the top may not be passed down to the bottom, especially when there is a change in management or personnel within the middle of the subcontracting tiers, then the message trail is lost, go blank already. Then how to cascade down to the bottom tier? The bottom tier would not know what the top wants?

Q25: So it can result in a mismatch of expectations and worse still, wrong specifications of the end product? Meaning end product is not what the client wants?

A25: Yes. Then there is a lot of wastage and abortive works.

Q26: Could you elaborate more on wastage and abortive works?

A26: Wastage are very clear on many things; materials and labour also known as resources. For example, you send your workers to site but never brief him. You don't know what you are supposed to do on site today and then you start to ask around on site. Then who will allow you to work, then chase you out and ask you to find your drawings and instructions. Hence, it is a waste of time and materials. You don't know, you don't plan and you order the materials and then the material cannot be used. That's another waste. When you waste this part, then you have lesser time to complete our work since you wasted one day and then you have a chain reaction where due to this work delay, you cannot go to another site.

Q27: There is this view that the multiple layers of subcontracting, so that the sub-cons, which are usually the specialist contractors, are able to complete the job more efficient. Is this right?

A27: This is just to cover their incompetency. If you want to be in this line, you must be qualified and competent to do it. You cannot rely on others to solve your problems without knowing what's going on. Singapore's construction landscape is a bit worrying now. People are using contracts to penalise subcontractors. You see if you are a big firm, you have all the resources and legal advice and documents. You know every single term in the SIA (Singapore Institute of Architects) contracts. So when the sub-con comes in, a lot of times they have to listen to the big boys even though they know you are bullying them. It's not healthy. The reason is in this industry and it has been in practice for a long time but I don't think it is a healthy trend for people who are practising it. Bottom-line is that the subcons are not very educated people, to a certain extent, they really don't know their rights. But to them, if life can go on, then they are very happy already. So they are being exploited. However, you must also look from the point that if sub-cons were to challenge the main-con by listing down their rights, then they will not get the job from the same main con ever again. It's best to have a main-con who is more open and the mind-set is willing to help the sub-cons.

Q28: How do you determine whether a construction firm should be classified as sub-con or main-con?

A28: I think it is more of the job requirement. You just look for the trade and then select a suitable firm. That's for sub-con or specialist contractors. For main-con, we go through a lot of recommendations, reputable company, by word of mouth etc. That's why there's a need of a platform to share information on these matters.

Q29: So basically, it is to go by trade and then by word of mouth to determine main-con or sub-con for construction firms. Am I right?

A29: Yes. At the end of the day, it boils down to dealing with people. Construction is all about people, it is a people business.

End of Verbatim Report 20

Bibliography

Barney, J. B. (1986). Organizational culture: Can it be a source of sustained competitive advantage? *Academy of Management Review, 11,* 656–665.

Barney, J. B. (1986). Strategic factor markets: Expectations, luck, and business strategy. *Management Science, 32,* 1231–1241.

Barney, J. B. (1991). Special theory forum: The resource-based model of the firm: Origins, implications, and prospects. *Journal of Management, 17,* 97–98.

Barney, J. B. (1996). The resource-based theory of the firm. *Organization Science, 7,* 469.

Barney, J. B. (1999). How a firm's capabilities affect boundary decisions? *Sloan Management Review, 40,* 137–145.

Barney, J. B. (2001). Is the resource-based "view" a useful perspective for strategic management research? Yes. *Academy of Management Review, 26,* 41–56.

Barney, J. B. (2005). Where does inequality come from? The personal and intellectual roots of resource-based theory. In K. G. Smith & M. A. Hitt (Eds.), *Great minds in management: The process of theory development* (pp. 280–303). Oxford, UK: Oxford University Press.

Barney, J. B., & Clark, D. N. (2007). *Resource-based theory: Creating and sustaining competitive advantage*. Oxford, UK: Oxford University Press.

Barney, J. B., & Hansen, M. H. (1994). Trustworthiness as a source of competitive advantage. *Strategic Management Journal, 15,* 175–190.

Barney, J. B., Wright, M., & Ketchen, D. J., Jr. (2001). The resource-based view of the firm: Ten years after 1991. *Journal of Management, 27,* 625–641.

Building and Construction Authority. (2012) Construction productivity roadmap, Singapore.

DOS. (2010). *Singapore standard industrial classification 2010*. Singapore: DOS. Retrieved February 9, 2015. Available online: http://www.singstat.gov.sg/methodologies_and_standards/standards_and_classifications/SSIC.html.

Grant, R. M. (1996). Prospering in dynamically-competitive environments: Organizational capability as knowledge integration. *Organization Science, 7,* 375–387.

Ng, H. (2011). *Relationship between risk management and project network in the Singapore construction industry* (Unpublished BSc. (Building) study). Singapore: National University of Singapore.

Ng, S. (2014). *Early contractor's involvement: A Contractor's perspective* (Unpublished BSc. (Building) study). Singapore: National University of Singapore.

Pettigrew, A., Thomas, H., & Whittington, R. (2002). Construction competitive advantage. In *Handbook of strategy and management*. London: Sage Publications.

Porter, M. E. (1979). How competitive forces shape strategy. *Harvard Business Review* 21–38.

Porter, M. E. (1979). How competitive forces shape strategy. *Harvard Business Review* 137–146.

L. Sui Pheng and S. Jia Zheng, *Construction Productivity in the Multilayer Subcontracting System*, Management in the Built Environment,
https://doi.org/10.1007/978-981-13-1831-3

Porter, M. E. (1979). The structure within industries and companies' performance. *Review of Economics and Statistics, 61*(2), 214–227.

Porter's Five Forces Strategy Skills. (2013). (1st ed., pp. 9). Retrieved from http://www.free-management-ebooks.com/dldebk/dlst-porter.htm.

Spanos, Y. E., & Lioukas, S. (2001). An examination into the causal logic of rent generation: Contrasting Porter's competitive strategy framework and the resource-based perspective. *Strategic Management Journal, 22*(10), 907–934.